Tony Evans

PROBLEMS OF
ATOMIC DYNAMICS

PROBLEMS OF
ATOMIC DYNAMICS

Max Born

The M.I.T. Press
Cambridge, Massachusetts, and London, England

Originally published in 1926 by
Massachusetts Institute of Technology

First M.I.T. Press Paperback Edition, October 1970

ISBN 0 262 52019 2

Library of Congress catalog card number: 74-123256

Printed in the United States of America

DEDICATED TO
PROFESSOR PAUL HEYMANS

FOREWORD

It has been the custom of the Massachusetts Institute of Technology to invite from time to time distinguished scientists and engineers to address its classes of students. The results of these occasional lectures have been so valuable that more systematic efforts have been made to secure the services of such lecturers as an adjunct to our scheme of instruction. It is now proposed to make the impression of these lectures deeper and more permanent by printing in book form wherever practicable the text or notes of such lectures. This publication of the text of lectures given during the winter of 1925–1926 by Professor Max Born of Göttingen is the first of the series.

C. L. NORTON

DEPARTMENT OF PHYSICS
MASSACHUSETTS INSTITUTE OF TECHNOLOGY
April, 1926

PREFACE

The lectures which constitute this book are given just as they were presented at the Massachusetts Institute of Technology from November 14, 1925, to January 22, 1926, without any amplification. They do not purport to be a text-book — for of these we have enough — but rather an exposition of the present status of research in those regions of physics in which I myself have made investigations, and of which I therefore believe that I can take a comprehensive view. In the short time that was at my disposal, I could neither seek for completeness nor consider minutiæ. It was my purpose to present methods, objects of investigation, and the most important results. I have avoided references and have only occasionally named individual authors. I take this occasion to ask the pardon of all those colleagues whose names I have omitted to mention.

The lectures on the theory of lattices are essentially an abstract of certain sections of my book, *Atomtheorie des festen Zustandes*, and of subsequent works on this topic. In the same manner, the earlier lectures on the structure of the atom are closely related to my book *Atommechanik*, but I soon made the transition to a different point of view. At the time I began this course of lectures, Heisenberg's first paper on the new quantum theory had just appeared. Here his masterly treatment gave the quantum theory an entirely new turn. The paper of Jordan and myself, in which we recognized the matrix calculus as the proper formulation for Heisenberg's ideas, was in press, and the manuscript of a third paper by the three of us was almost completed. Though the results contained in this third paper left no doubt in my mind as to the superiority of the new methods to the old, I could not bring myself to plunge directly into the new quantum mechanics. To do this would not only be to deny to Bohr's great achievement its due need of credit, but even more to deprive the reader of the natural and marvelous development of an idea. I have consequently begun by presenting the Bohr theory as an application of classical mechanics, but have emphasized more than is usual its weaknesses and conceptual difficulties. It is perhaps superfluous to state that this is only done to establish the necessity of a new conception, and is not intended as a hostile criticism of Bohr's immortal work. As the course proceeded, further achievements of the new method came to my notice. I was able to introduce some of these into the lectures. Pauli's theory of the hydrogen atom is a case in point. Of others, such as the treatment of the theory of aperiodic processes in terms of a general calculus of operators, developed by N. Wiener and myself, I was able to give a sketch. These sections are not so much a report on scientific results as an enumeration of the problems which seem of most interest to us theoretical physicists.

I wrote the original text in German. It was then translated into English by Dr. W. P. Allis and Mr. Hans Müller, and read through by me. Mr. F. W. Sears revised the second part; finally Dr. M. S. Vallarta went carefully through the complete text in order to verify the formulas and make the English idiomatic. I hereby express my sincere thanks to all these gentlemen, who have spent much labor on this work of revision, and have sacrificed much valuable time, as well as to Assistant Dean H. E. Lobdell, who has taken great pains in the supervision of the work of publication.

I feel it as a great honor that this book appears as a publication of the Massachusetts Institute of Technology. For this I wish to express my thanks to President S. W. Stratton and Professor C. L. Norton, the Head of the Department of Physics. To Professor Paul Heymans, who has not merely extended to my wife and myself the hospitality of his house, but also shared his office with me during the three months of my stay at the Institute, I wish to express my gratitude in visible form by the dedication of this little book.

MAX BORN

Massachusetts Institute of Technology
January, 1926

CONTENTS

	PAGE
FOREWORD	vii
PREFACE	ix

SERIES I

THE STRUCTURE OF THE ATOM

LECTURE 1 1
    Comparison between the classical continuum theory and the quantum theory — Chief experimental results on the structure of the atom — General principles of the quantum theory — Examples.

LECTURE 2 12
    General introduction to mechanics — Canonical equations and canonical transformations.

LECTURE 3 21
    The Hamilton-Jacobi partial differential equation — Action and angle variables — The quantum conditions.

LECTURE 4 25
    Adiabatic invariants — The principle of correspondence.

LECTURE 5 32
    Degenerate systems — Secular perturbations — The quantum integrals.

LECTURE 6 38
    Bohr's theory of the hydrogen atom — Relativity effect and fine structure — Stark and Zeeman effects.

LECTURE 7 47
    Attempts towards a theory of the helium atom and reasons for their failure — Bohr's semi-empirical theory of the struc-

ture of higher atoms — The optical electron and the Rydberg-Ritz formula for spectral series — The classification of series — The main quantum numbers of the alkali atoms in the unexcited state.

LECTURE 8 54

Bohr's principle of successive building of atoms — Arc and spark spectra — X-ray spectra — Bohr's table of the completed numbers of electrons in the stationary states.

LECTURE 9 60

Sommerfeld's inner quantum numbers — Attempts toward their interpretation by means of the atomic angular momentum — Breakdown of the classical theory — Formal interpretation of spectral regularities — Stoner's definition of subgroups in the periodic system — Pauli's introduction of four quantum numbers for the electron — Pauli's principle of unequal quantum numbers — Report on the development of the formal theory.

LECTURE 10 68

Introduction to the new quantum theory — Representation of a coördinate by a matrix — The elementary rules of matrix calculus.

LECTURE 11 75

The commutation rule and its justification by a correspondence consideration — Matrix functions and their differentiation with respect to matrix arguments.

LECTURE 12 79

The canonical equations of mechanics — Proof of the conservation of energy and of the "frequency condition" — Canonical transformations — The analogue of the Hamilton-Jacobi differential equation.

LECTURE 13 83

The example of the harmonic oscillator — Perturbation theory.

LECTURE 14 89

The meaning of external forces in the quantum theory and corresponding perturbation formulas — Their application to the theory of dispersion.

CONTENTS

	PAGE
LECTURE 15	94

Systems of more than one degree of freedom — The commutation rules — The analogue of the Hamilton-Jacobi theory — Degenerate systems.

LECTURE 16 99

Conservation of angular momentum — Axial symmetrical systems and the quantization of the axial component of angular momentum.

LECTURE 17 106

Free systems as limiting cases of axially symmetrical systems — Quantization of the total angular momentum — Comparison with the theory of directional quantization — Intensities of the Zeeman components of a spectral line — Remarks on the theory of Zeeman separation.

LECTURE 18 113

Pauli's theory of the hydrogen atom.

LECTURE 19 119

Connection with the theory of Hermitian forms — Aperiodic motions and continuous spectra.

LECTURE 20 125

Substitution of the matrix calculus by the general operational calculus for improved treatment of aperiodic motions — Concluding remarks.

SERIES II

THE LATTICE THEORY OF RIGID BODIES

LECTURE 1 133

Classification of crystal properties — Continuum and lattice theories — Geometry of lattices.

LECTURE 2 139

Molecular forces — Polarizability of atoms — Potential energy and inner forces — Homogeneous displacements — The conditions of equilibrium — Examples of regular lattices.

CONTENTS

LECTURE 3 146

Elimination of inner motions — Compressibility — Elasticity and Hooke's law — Cauchy's relations — Dielectric displacement and piezoelectricity — Residual-ray frequencies.

LECTURE 4 155

Ionic lattices — Kossel's and Lewis' theory — Calculation of the lattice energy according to Madelung and Ewald.

LECTURE 5 163

The energy of the rock-salt lattice — Repulsive forces — Derivation of the properties of salt crystals from the properties of inert gases.

LECTURE 6 168

Experimental determination of the lattice energy by means of cyclic processes — The electron affinity of halogens — Heat of dissociation of salt molecules — Theory of molecular structure.

LECTURE 7 176

Chemical crystallography — Coördination lattices — Hund's theory of lattice types — Molecule, radical and layer lattices.

LECTURE 8 183

Physical mineralogy — The parameters of asymmetrical lattices — The molecule lattice of hydrochloric acid — Bragg's calculation of the rhombohedral angle of calcite — Rutile and anatase — Influence of the polarizability on elastic and electric constants — The breaking stress of rock salt.

LECTURE 9 189

Crystal optics — Refraction and double refraction — Optical activity — Thermodynamics — Quantum theory of specific heats — Distribution of frequencies in phase space.

LECTURE 10 196

Thermal expansion and pyroelectricity — Concluding remarks.

SERIES I
THE STRUCTURE OF THE ATOM

SERIES I
THE STRUCTURE OF THE ATOM

LECTURE 1

Comparison between the classical continuum theory and the quantum theory — Chief experimental results on the structure of the atom — General principles of the quantum theory — Examples.

Physics today is everywhere based on the theory of atoms. Through experimental and theoretical researches we have reached the conviction that matter is not infinitely divisible, but that there exist ultimate units of matter which cannot be further divided. However, it is not the atoms of the chemists that we feel authorized in calling "indivisible"; on the contrary they are very complicated structures composed of smaller elements. These are, from the point of view of recent investigations, the atoms of electricity, the (negative) electrons and the (positive) protons. It is conceivable that at a later epoch science will change its point of view and penetrate to still smaller elements; in this case the philosophical significance of atomistics could no longer be valued as highly. The last units would not be anything absolute, but only a measure of the present status of science. But I do not think that is so; I believe that we can hope that we have not to do with an endless chain of divisions, but that we are near the end of a finite chain, perhaps we have even attained it. The reasons that can be given for this optimism lie less in the experimental evidence for the reality of atoms, protons and electrons, which the new physics has furnished, than in the special character of the laws which govern the interactions of elementary electric particles. These laws have indeed properties which permit us to conclude that we are near their final formulation.

Such an assertion may seem too bold, because all philosophies of all ages have taught that human knowledge is incomplete, that each goal of knowledge is attained only at the cost of new puzzles. Up to the present, in physics as in other sciences, every result that our age has proclaimed as absolute has had to fall after a few years, decades or centuries, because new investigations have brought new knowledge and we have become used to consider the true laws of nature as unattainable ideals to which the so-called laws of physics are only successive approximations. Now, when I say that certain formulations of the laws of the atomistics of today have a character which is in a certain sense final, this does not fit in with our scheme of successive approximations and it becomes necessary that I offer an explanation. This special character that the atom possesses is the appearance of *whole numbers*. We pretend not only that in any body, for instance a piece of metal, there exist a certain finite number of atoms or electrons, but further that the properties of a single atom and the processes which occur during the interaction of several atoms are capable of being described by whole numbers. This is the substance of the *quantum theory*, the fundamental significance of which is based not only in its practical application but above all in its philosophical consequences considered here. To illustrate this idea we consider a small body free to move in a straight line. According to the usual ideas it can be at any time at any point. To fix this point we give the coördinate x measured from a point 0.

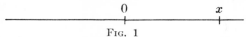
Fig. 1

But the accuracy of this indication depends entirely on experimental means of observation. If x can vary continuously, a more exact measurement may give us another decimal. For the processes in the atom, however, conditions seem to be different. We may compare them with the behavior of this body if we consider it infinitely small and allow it to occupy only certain discrete points which we shall number 1, 2, 3, · · · · . The coördinate x can therefore only take the values 1, 2, 3, · · · but not, for instance, $\frac{1}{2}$ or 3.7. This is in fact the behavior of

THE STRUCTURE OF THE ATOM

the so-called quantum numbers by means of which we describe today the state of atoms. Should this process be always satisfactory we evidently stand before a new state of our knowledge. If the value of x should be exactly a whole number, then a determination of such a number once made could not be altered. If it had been determined that x is certainly not equal to 1, nor to 3 or 4, nor to any greater number, then there remains for x only the value 2, and a more accurate measurement cannot change anything. We have therefore definite elements in the statements of laws, and there seems to exist a tendency that laws obtain this essential final character when expressed as relations between whole numbers. I therefore do not exaggerate when I say that the year 1900, when Planck first stated his theory of quanta, marked the beginning of an entirely new conception of nature.

The theory of matter, as treated up to the present, still falls very short of this extreme view. To emphasize this standpoint, we consider again the body in the straight line with the coördinate x. Then the usual quantum theory corresponds somewhat to the condition that x be allowed to take all possible continuous values, but that then the integral values of x should be selected as stationary states through the so-called quantum conditions. This conception is altogether too unsatisfactory. For this reason we have sought at Göttingen to find a new formulation of the quantum theory in which only these integral values of x occur and intermediate fractional values have no meaning. This theory has been verified in the sense that certain fundamental difficulties that existed in the old quantum theory are not encountered in the new. On the other hand, calculations are rather complicated. Therefore I shall not begin my course of lectures with this new theory, but start with a short survey of the old theory. Let me remind you of the most important experimental investigations of the structure of atoms.

The first of these is the conception developed by Lenard and Rutherford, that the atom is composed of a positive *nucleus* surrounded by negative *electrons*. The simplest atom, that of hydrogen, consists of one electron revolving around the simplest nucleus, a *proton*, each having the same charge $e = 4.77 \times 10^{-10}$ e.s.u., but different masses, the ratio of the masses being 1 : 1830. The nuclei of the other atoms are complex structures built up of protons and electrons, as shown by radioactive phenomena, but in these lectures we shall not discuss the structure of these nuclei, but treat them as masspoints with a charge which is an integral multiple Z of the charge e given above. This number Z is known as the atomic number and determines the position of the element in the periodic system. In the neutral atoms the number of electrons is also Z; in negative ions the number of electrons is greater than Z, in positive ions it is less.

The forces binding the electrons to the nucleus are certainly of electrical nature. This has been proved by the experiments of Lenard on the scattering of cathode rays and the experiments of Rutherford and his students on the scattering of α-rays, in which it was shown that Coulomb's law of force holds for distances of the order of magnitude involved in this theory.

But the supposition of purely electrical forces leads to difficulties. There is a mathematical theorem which states that a system of electric charges cannot be in stable equilibrium; therefore Rutherford was compelled to assume that the electrons move around the nucleus in such a way that the centrifugal force balances the resultant of the electrical forces. But if electromagnetic laws can be applied to such a system, it must radiate energy until the electrons fall into the nucleus. A second difficulty arises from the kinetic theory of gases. We know that every molecule or atom of a gas under normal conditions collides with other molecules or atoms about 100,000,000 times per second. If the ordinary laws of mechanics held, there would be expected a slight change of the electron orbits at each collision and these changes would accumulate so that after one second the system would be materially altered. But we know that every molecule has a definite set of properties. It is therefore necessary to find a principle of stability which evidently cannot be derived from the ordinary laws of mechanics.

Niels Bohr has given this principle by applying the rules of quantum theory to atomic systems. These rules were devel-

THE STRUCTURE OF THE ATOM

oped by Max Planck in the study of the laws of heat radiation. He proved that it is impossible to explain the spectral distribution of the energy radiated by a black body if we make the ordinary assumption that energy can be divided into infinitely small parts; but it may be explained if we assume that the energy exists in quanta of finite size, $h\nu$, where ν is the frequency of the radiation and h is a constant, $h = 6.54 \times 10^{-27}$ erg. sec. This remarkable idea has been of the greatest fruitfulness in the development of physics, for it has been shown that the constant h and the quantum $h\nu$ play important rôles in many phenomena. In the photoelectric effect the kinetic energy of the photoelectron is given by $mv^2/2 = h\nu$, where ν is the frequency of the incident radiation. This equation, proposed by Einstein, was proved experimentally by Millikan and others and gave the first direct evidence of the existence of the quantum. It was followed by many other experiments of a similar kind of which I will mention only one group: that investigating the relation between the kinetic energy of an electron and the frequency of the light emitted as the result of the collision between this electron and an atom, first tested by Franck and Hertz and later developed by Compton, Foote, Mohler and many other American physicists.

All these experiments show that the production of radiation of a certain frequency requires a certain amount of kinetic energy. Niels Bohr has made the assumption that this law holds not only between kinetic energy and radiation, but between all kinds of energy and radiation. In this way he found a very simple interpretation of the fact that isolated atoms, as in a rarefied gas, emit a line spectrum, that is, a set of monochromatic light waves. He assumes that Einstein's law can be applied to the emission of a line in the spectrum in such a way that while the system loses a finite amount of internal energy $W_1 - W_2$, the frequency ν of the emitted light is connected with this loss of energy by the equation

$$W_1 - W_2 = h\nu. \qquad (1)$$

To explain the whole system of lines of the atom Bohr postulated the existence of a system of so-called "stationary states" in which the atom can exist without loss of energy by radiation, while keeping its total energy contents $W_1, W_2 \cdots$. The frequency of every spectral line appears now as the difference of two terms W_1/h and W_2/h, in perfect agreement with the well-known optical fact formulated in the Ritz combination principle. At the same time this hypothesis solves the difficulty previously mentioned concerning the stability of atomic systems, for the energy necessary to change an atom from one stationary state to another is large, larger than that available at ordinary temperatures as a consequence of thermal agitation, therefore the atom remains unaltered.

These assumptions are in direct contradiction to classical dynamics, but lacking any knowledge of the exact laws of the new theory we use the classical laws as far as possible and then seek to alter them, when they lead nowhere. The chief problem becomes the determination of the stationary states and their energies; but first we show, after Einstein, that Bohr's principle suffices to give a very simple derivation of Planck's formula for *black-body radiation*.

Consider two stationary states W_1 and W_2 ($W_1 > W_2$). In statistical equilibrium they may exist in the amounts N_1 and N_2. Then by Boltzmann's principle

$$\frac{N_2}{N_1} = \frac{e^{-W_2/kT}}{e^{-W_1/kT}} = e^{\frac{W_1-W_2}{kT}}$$

and using Bohr's frequency condition (1):

$$\frac{N_2}{N_1} = e^{\frac{h\nu}{kT}}$$

In the classical theory, the interaction of atomic systems and radiation is made up of three processes:

1. If the atom is in a state of higher energy it loses energy spontaneously by outward radiation.
2. The external radiation field adds or subtracts energy to or from the atom depending on the phase and amplitude of the waves of which it consists. We call these processes:

(a) positive absorption if the atom gains energy,
(b) negative absorption if the atom loses energy through the action of the external field.

In the last two cases the contribution of these processes to the change of energy is proportional to the energy density ρ_ν.

In analogy to these we assume for the quantic interaction three corresponding processes. The following transitions occur between the two energy levels W_1 and W_2:

1. Spontaneous decrease in energy through changes from W_1 to W_2. The frequency with which these transitions occur is proportional to the number of systems, N_1, which are in the initial state W_1 and is also dependent on the final state. For the number of these transitions we therefore write,

$$A_{12}N_1.$$

2a. Increases in energy due to the radiation field (transitions from W_2 to W_1). We place, likewise, for the number of such transitions

$$B_{21}N_2\rho_\nu.$$

2b. Decreases in energy due to the radiation field (transitions from W_1 to W_2). The number of such transitions is

$$B_{12}N_1\rho_\nu.$$

For statistical equilibrium between the states W_1 and W_2 it is required that

$$A_{12}N_1 = (B_{21}N_2 - B_{12}N_1)\rho_\nu,$$

from which

$$\rho_\nu = \frac{A_{12}}{B_{21}\dfrac{N_2}{N_1} - B_{12}} = \frac{A_{12}}{B_{21}e^{\frac{h\nu}{kT}} - B_{12}}. \tag{2}$$

It is natural to suppose that the classical laws are limiting cases of the quantum laws. Here the limiting case is that of high temperatures, where $h\nu$ is small compared with kT. Under such conditions Equation (2) should go over into the classical law of Rayleigh and Jeans,

$$\rho_\nu = \frac{8\pi}{c^3} \nu^2 kT.$$

For large values of T (2) has the form

$$\rho_\nu = \frac{A_{12}}{B_{21} - B_{12} + B_{21}\frac{h\nu}{kT} + \cdots}.$$

These two expressions become identical if

$$B_{12} = B_{21},$$

and

$$\frac{A_{12}}{B_{12}} = \frac{8\pi}{c^3} \nu^3 h.$$

Inserting these values in (2) we obtain Planck's radiation formula

$$\rho_\nu = \frac{8\pi h}{c^3} \frac{\nu^3}{e^{\frac{h\nu}{kT}} - 1}.$$

We see that the validity of Planck's formula is quite independent of the determination of the stationary states.

We shall now consider the problem of the *determination of the stationary states*. The simplest model of a radiating system is the *harmonic oscillator*, the equation of motion of which is

$$m\ddot{q} + \kappa q = 0$$

where q is the distance of the moving point from the position of equilibrium, m its mass, and κ a constant which is connected with the natural frequency ν_0 by the relation

$$\kappa = m(2\pi\nu_0)^2.$$

The motion of a point obeying this equation is very closely related to the motion of the field vector in a monochromatic light wave. An immediate assumption is that the frequency of such a linear oscillator is the same as the frequency of the emitted light; then it follows from Bohr's frequency condition (1) that the energies of the stationary states of the oscillator

THE STRUCTURE OF THE ATOM

must differ by $h\nu_0$, that is they are, for an appropriate choice of additive constants,

$$W_0 = 0, \quad W_1 = h\nu_0, \quad W_2 = 2h\nu_0 \cdots, \quad W_n = nh\nu_0 \cdots.$$

In the case of one degree of freedom the motion is completely determined by the energy, so in this simple example the stationary states are completely known,

$$q = \sqrt{\frac{W}{2\pi m\nu_0^2}} \cos(2\pi\nu_0 t + \delta).$$

From the system of energy levels it is possible to derive the complete system of spectral lines by taking all possible differences,

$$\nu = \frac{1}{h}(nh\nu_0 - kh\nu_0) = \nu_0(n-k).$$

We see that Bohr's principle gives for the frequency of the emitted light not only the fundamental frequency, ν_0, as in the classical theory, but also the overtones $\nu_0(n-k)$. But in this simple case of the linear oscillator we should expect that both theories give exactly the same result; therefore we need a new principle to eliminate superfluous overtones. Bohr has supplied this under the name of the *Principle of Correspondence*. He makes the assumption we have already used once, that the quantum laws must go over into the classical laws in the limiting case. If the oscillator has a very large amount of energy, that is if n is large, the difference between two neighboring energy-levels is small compared with their absolute values, and approximately the series of values of W_n may be looked upon as varying continuously, as in the classical theory. Therefore Bohr assumes that the classical theory remains approximately valid in this limiting case. Then the emitted light may be calculated classically, and the light vector is proportional to the electric moment of the vibrating system. In the case of one coördinate q, this moment is eq, where e is the charge of the moving point, and for the oscillator we have,

$$eq = e\sqrt{\frac{W}{2\pi m\nu_0^2}} \cos(2\pi\nu_0 t + \delta).$$

In the general case the electric moment is a Fourier series with an infinite number of terms of the same form. The squares of the coefficients of the terms of the series will be a measure of the intensity of light emitted at the corresponding frequency of the overtone; this measure must also hold approximately in the quantum theory. In this way we get a rough estimate of the intensities of spectral lines even for small quantum numbers. But in one case we may expect this rule to give the exact results, namely, when one Fourier coefficient is identically zero. Then we may assume that the corresponding frequency is not emitted at all and that the corresponding transition does not occur. In our case we see that the Fourier series has only one term, corresponding to the transition $n - k = 1$. In this way Bohr's correspondence principle reduces the number of frequencies to that of the classical theory. It seems quite trivial in this example, but we shall see that in other cases it gives valuable information concerning possible transitions.

We go over now to more complicated systems, at first those of one degree of freedom but of any energy-function. In a system of this kind we have in general, if we exclude orbits which go to infinity, periodic motions, so that the coördinate q can be expanded in a Fourier series of the time t. It might be thought that it would be possible to determine the stationary states, as in the case of the oscillator, by making the energies integral multiples of $h\nu$, where $T = 1/\nu$ is the period of the motion, i.e., the time of a complete revolution. We shall see that this is not possible on account of another principle, the last to be mentioned in this introduction: the *adiabatic hypothesis* of Ehrenfest.

We consider the action of an external force on the atomic system. There are two limiting cases: constant forces and oscillating forces of high frequency. We know that in the second case classical mechanics cannot be applied, for the action of light consists in the production of transitions or quantum jumps which cannot be described by classical theory. But suppose that we investigate the action of a force changing slowly relatively to the inner motions. From the assumption of stationary states it follows that this force can have either no effect or a finite effect resulting in a quantum jump; it is natural to suppose that the latter case will be more and more improbable as the rate of change of the force diminishes. So we see that quantities which are suitable for fixing the stationary states must have the property of not being altered by slowly changing forces. Ehrenfest calls them "adiabatic invariants," from analogy to similar quantities in thermodynamics. The question is whether such quantities can be found at all in classical mechanics. In the case of a linear oscillator it is not the energy which has this property, because the frequency ν is not constant under the influence of slowly changing forces, but it can be proved that the quotient W/ν is an adiabatic invariant. Indeed our fixation of the stationary states of the oscillator can be formulated by giving to this quotient the discrete values $0\,h$, $1\,h$, $2\,h\cdots$ etc. One of the aims of these lectures will be to find adiabatic invariants for every atomic system; we shall show that they exist not only for simple periodic systems, but also for the larger class of so-called multiple-periodic systems.

Periodic properties are closely connected with the laws of the quantum theory. A process which can be resolved into rotations and oscillations falls in its domain. Therefore, it will be our first problem to study systematically the most general systems having periodic properties.

LECTURE 2

General introduction to mechanics — Canonical equations and canonical transformations.

The laws of mechanics can be best condensed, by means of the Principle of Least Action, in Hamilton's formula,

$$\int_{t_1}^{t_2} L\,dt \text{ is stationary.} \tag{1}$$

In this formula, the so-called Lagrangian function L depends on the coördinates and the components of the velocities and the extremal is understood to be taken by comparison of all motions leading from one given point Q_1 at the time t_1 to another given point Q_2 at the time t_2. This formulation has the advantage that it is independent of the system of coördinates. In the following we shall always operate with generalized independent coördinates $q_1, q_2 \cdots$. Using the variational principle we obtain, from the variation of equation (1), the equations of motion in the form of Euler-Lagrange

$$\frac{d}{dt}\frac{\partial L}{\partial \dot{q}_k} - \frac{\partial L}{\partial q_k} = 0. \tag{2}$$

We shall now give the expression for L in three important cases:
1. In the mechanics of Galileo and Newton,

$$L = T - U$$

where T is the kinetic and U the potential energy. Denoting the velocity vectors by $\mathbf{v}_k$ and the masses by m_k we have

$$T = \tfrac{1}{2} \sum_k m_k \mathbf{v}_k^2$$

and Equations (2) take the Newtonian form,

$$\frac{d}{dt}(m_k \mathbf{v}_k) = \mathbf{F}_k \tag{3}$$

THE STRUCTURE OF THE ATOM 13

where the components of the forces $\mathbf{F}_k$ are obtained from U by differentiation.

$$F_{kx} = -\frac{\partial U}{\partial x_k} \cdot \cdot \cdot \cdot$$

2. In Einstein's relativistic mechanics we have

$$L = T^\times - U$$

where

$$T^\times = \sum_k m_k^0 c^2 (1 - \sqrt{1 - (v_k/c)^2}) \tag{4}$$

and m_k^0 is the rest-mass, v_k the magnitude of the velocity, and c the velocity of light. This $T^\times$ is *different* from the kinetic energy

$$T = \sum_k m_k^0 c^2 \left(\frac{1}{\sqrt{1 - (v_k/c)^2}} - 1 \right). \tag{5}$$

In this case the equations of motion can also be written in the form (3) if the mass depends on the velocity according to the law:

$$m_k = \frac{m_k^0}{\sqrt{1 - \left(\frac{v_k}{c}\right)^2}}. \tag{6}$$

3. If magnetic forces are acting on the system we have,

$$L = T - U - \frac{1}{c} \sum_k e_k \mathbf{A}_k \cdot \mathbf{v}_k \tag{7}$$

where e_k is the charge of the particle and $\mathbf{A}_k$ the vector potential of the magnetic field for that configuration of the system.

The equations of motion are of the second order with respect to time and it is often convenient to transform them into twice as many equations of the first order. This has been done by Hamilton in the following symmetrical way:

Introducing as unknown functions besides the coördinates the momenta

$$p_k = \frac{\partial L}{\partial \dot{q}_k} \tag{8}$$

and using instead of $L(q_1, \dot{q}_1, q_2, \dot{q}_2 \cdots)$ the function

$$H(q_1, p_1, q_2, p_2 \cdots) = \sum_k p_k \dot{q}_k - L \tag{9}$$

the Principle of Least Action can be written

$$\int_{t_1}^{t_2} [\sum_k p_k \dot{q}_k - H(q_1, p_1 \cdots)]dt \text{ is stationary} \tag{10}$$

and the equations of Euler-Lagrange take the symmetrical form

$$\begin{cases} \dot{q}_k = \dfrac{\partial H}{\partial p_k} \\ \dot{p}_k = -\dfrac{\partial H}{\partial q_k} \end{cases} \tag{11}$$

Equations (11) are also true if the Hamilton function H depends explicitly on the time t. If this is *not* the case we find

$$\frac{dH}{dt} = \sum_k \left(\frac{\partial H}{\partial q_k} \dot{q}_k + \frac{\partial H}{\partial p_k} \dot{p}_k \right) = 0$$

or

$$H = \text{const.} \tag{12}$$

We shall now discuss the physical meaning of H in the same three cases considered above:

1. In the mechanics of Galileo and Newton, where T is a homogeneous quadratic function of the components of the velocities we have, according to Euler's theorem,

$$2T = \sum_k \frac{\partial T}{\partial \dot{q}_k} \dot{q}_k = \sum_k \frac{\partial L}{\partial \dot{q}_k} \dot{q}_k = \sum_k p_k \dot{q}_k$$

and therefore, according to (9) and since $L = T - U$,

$$H = T + U.$$

H is therefore the total energy and (12) is the law of conservation of energy. This holds only for "inertial systems" and not for accelerated systems of coördinates. In such a system, for instance in a rotating system, H is constant, but does not represent the energy.

THE STRUCTURE OF THE ATOM 15

2. In relativistic mechanics we find by a simple calculation

$$H = \sum_k m_k^0 c^2 \left(\frac{1}{\sqrt{1 - \left(\frac{v_k}{c}\right)^2}} - 1 \right) + U = T + U$$

whence H is also the total energy. If it is desired to express this in terms of the momenta, we find, combining the momenta corresponding to the components of velocity with a vector momentum

$$\mathbf{p}_k = m_k \mathbf{v}_k = \frac{m_k^0 \mathbf{v}_k}{\sqrt{1 - \left(\frac{v_k}{c}\right)^2}}$$

and by elimination of $\mathbf{v}_k$

$$H = \sum_k m_k^0 c^2 \left(\sqrt{1 + \frac{\mathbf{p}_k^2}{(m_k^0)^2 c^2}} - 1 \right) + U. \tag{13}$$

3. In a magnetic field we do not have a simple proportionality between velocity and momentum, but

$$\mathbf{p}_k = m_k \mathbf{v}_k - \frac{e_k}{c} \mathbf{A}_k$$

and even in this case H is the total energy

$$H = T + U.$$

Introducing the momenta we find a rather complicated expression

$$H = \sum_k \left(\frac{\mathbf{p}_k^2}{2 m_k} + \frac{e_k}{c m_k} \mathbf{A}_k \cdot \mathbf{p}_k + \frac{e_k^2}{2 m_k c^2} \mathbf{A}_k \cdot \mathbf{A}_k \right) + U. \tag{14}$$

Before beginning with the general integration of the canonical equations, we shall consider some simple examples. If the Hamiltonian function, H, is independent of one coördinate, for example of q_1,

$$H = H(p_1, q_2, p_2 \cdots t),$$

we get from the canonical equations
$$\dot{p}_1 = 0$$
therefore,
$$p_1 = \text{const.},$$
and so we have found an integral of these equations. Such is the case, for example, if q_1 is the angle of rotation about an axis passing through the center of gravity of a solid body, therefore the coördinate is called a "cyclic" variable. In this case it is easily shown that p_1 is the moment of momentum of the system about the axis.

It may happen that H is independent of all the q_k's
$$H(p_1 \, p_2 \cdots t)$$
then the canonical equations are completely integrated by the formulas

$$\dot{p}_k = 0 \qquad p_k = \alpha_k$$
$$\dot{q}_k = \frac{\partial H}{\partial p_k} = \omega_k \qquad q_k = \omega_k t + \beta_k \qquad (15)$$

where the ω_k's are characteristic constants of the system, α_k and β_k constants of integration.

We see that the mechanical problem is solved if we can find such coördinates that H depends only on the momenta. This is the method of integration which we shall use in the following. The difficulty now is that variables of this type cannot be found by means of a simple point transformation of q_k, but only by a simultaneous transformation of q_k and p_k.

We shall now find all the transformations of p_k and q_k which do not change the form of the canonical equations. Such transformations are called "canonical transformations." This condition is evidently fulfilled if the Principle of Least Action (1) does not change its form by a transformation

$$p_k = p_k(\bar{q}_1, \bar{q}_2, \cdots \bar{p}_1, \bar{p}_2, \cdots t)$$

and

$$q_k = q_k(\bar{q}_1, \bar{q}_2, \cdots \bar{p}_1, \bar{p}_2, \cdots t);$$

THE STRUCTURE OF THE ATOM 17

in other words, if the sum

$$\sum_k p_k \dot{q}_k - H(q_1, p_1 \cdots t)$$

differs from the corresponding expression in the new coördinates by a quantity which is a total differential of the time. We must set, therefore,

$$\left[\sum_k p_k \dot{q}_k - H(q_1, p_1 \cdots t)\right] - \left[\sum_k \bar{p}_k \dot{\bar{q}}_k - H(\bar{q}_1, \bar{p}_1 \cdots t)\right] = \frac{dV}{dt}. \quad (16)$$

This equation is easily satisfied. Let us choose for V an arbitrary function of the new and old coördinates and of the time

$$V(q_1, \bar{q}_1, \cdots t).$$

We obtain by comparison of the coefficients of $\dot{q}_k$ and $\dot{\bar{q}}_k$:

$$\begin{cases} p_k = \dfrac{\partial}{\partial q_k} V(q_1, \bar{q}_1, \cdots t) \\ \bar{p}_k = -\dfrac{\partial}{\partial \bar{q}_k} V(q_1, \bar{q}_1, \cdots t) \\ H = \bar{H} - \dfrac{\partial}{\partial t} V(q_1, \bar{q}_1, \cdots t). \end{cases} \quad (17)$$

Expressing $\bar{q}_k$, $\bar{p}_k$ in terms of q_k, p_k we obtain the desired equations of transformation. But we can give to these canonical transformations several other forms, by using, instead of q_k, $\bar{q}_k$, other independent variables. There are in all four such combinations possible from which we select the common case where q_k, $\bar{p}_k$ are used as independent variables. To do this we write instead of V

$$V - \sum_k \bar{p}_k \bar{q}_k$$

which is evidently, like V, an arbitrary function, and consider here V as a function of q_k, $\bar{p}_k$. Then we obtain

$$[\sum_k p_k \dot{q}_k - H(q_1, p_1 \cdots t)] - [-\sum_k \bar{q}_k \dot{\bar{p}}_k - \bar{H}(\bar{q}_1, \bar{p}_1 \cdots t)]$$

$$= \frac{d}{dt} V(q_1, \bar{p}_1 \cdots t)$$

and therefore, by comparison of the coefficients,

$$\begin{cases} p_k = \dfrac{\partial}{\partial q_k} V(q_1, \bar{p}_1 \cdots t) \\ \bar{q}_k = \dfrac{\partial}{\partial \bar{p}_k} V(q_1, \bar{p}_1 \cdots t) \\ H = \bar{H} - \dfrac{\partial}{\partial t} V(q_1, \bar{p}_1 \cdots t). \end{cases} \qquad (18)$$

We illustrate this equation by a few simple examples:
The function
$$V = q_1 \bar{p}_1 + q_2 \bar{p}_2$$
gives the identical transformation
$$q_1 = \bar{q}_1, \qquad p_1 = \bar{p}_1, \qquad q_2 = \bar{q}_2, \qquad p_2 = \bar{p}_2.$$
The function
$$V = q_1 \bar{p}_1 \pm q_1 \bar{p}_2 + q_2 \bar{p}_2$$
gives
$$\begin{cases} q_1 = \bar{q}_1 \\ q_2 = \bar{q}_2 \pm \bar{q}_1 \end{cases} \qquad \begin{aligned} p_1 &= \bar{p}_1 \pm \bar{p}_2 \\ p_2 &= \bar{p}_2. \end{aligned}$$

For three pairs of variables the function
$$V = q_1(\bar{p}_1 + \bar{p}_2 + \bar{p}_3) + q_2(\bar{p}_1 + \bar{p}_3) + q_3 \bar{p}_3$$
gives the transformation
$$\begin{aligned} q_1 &= \bar{q}_1 & p_1 &= \bar{p}_1 + \bar{p}_2 + \bar{p}_3 \\ q_2 &= \bar{q}_2 - \bar{q}_1 & p_2 &= \bar{p}_2 + \bar{p}_3 \\ q_3 &= \bar{q}_3 - \bar{q}_2 & p_3 &= \bar{p}_3. \end{aligned}$$

In these examples the coördinates and impulses are transformed among themselves. The general condition is that V shall be a linear function of q and $\bar{p}$
$$V = \sum_{i,k} \alpha_{ik} q_i \bar{p}_k + \sum_k \beta_k q_k + \sum_k \gamma_k \bar{p}_k.$$
Then we have
$$p_i = \sum_k \alpha_{ik} \bar{p}_k + \beta_k$$
$$\bar{q}_i = \sum_k \alpha_{ki} q_k + \gamma_i.$$

THE STRUCTURE OF THE ATOM

If β_i and γ_i vanish we have

$$\sum_k p_k q_k = \sum_{k,l} \alpha_{kl} \bar{p}_l q_k = \sum_l \bar{q}_l \bar{p}_l.$$

This transformation is linear, homogeneous and contragredient. To this group belongs the case of orthogonal transformations, for instance the rotation of rectangular coördinates. We obtain a point transformation, that is a transformation of the q_k's among themselves, when V is linear in $\bar{p}$:

$$V = \sum_k f_k(q_1, q_2 \cdots) \bar{p}_k + g(q_1, q_2 \cdots)$$

that is

$$\begin{cases} p_k = \sum_l \dfrac{\partial f_l}{\partial q_k} \bar{p}_l + \dfrac{\partial g}{\partial q_k} \\ \bar{q}_k = f_k(q_1, q_2 \cdots) \end{cases}$$

and we have corresponding relations for the momenta.

As an example we shall give the transformation of rectangular coördinates into polar coördinates. Here we place,

$$-V = p_x r \cos\phi \sin\theta + p_y r \sin\phi \sin\theta + p_z r \cos\theta.$$

Then we have

$$\begin{cases} x = r \cos\phi \sin\theta & p_r = p_x \cos\phi \sin\theta + p_y \sin\phi \sin\theta + p_z \cos\theta \\ y = r \sin\phi \sin\theta & p_\phi = -p_x r \sin\phi \sin\theta + p_y r \cos\phi \sin\theta \\ z = r \cos\theta & p_\theta = p_x r \cos\phi \cos\theta + p_y r \sin\phi \cos\theta \\ & \quad - p_z r \sin\theta. \end{cases}$$

and the expression $p_x^2 + p_y^2 + p_z^2$ is transformed into

$$p_r^2 + \frac{1}{r^2} p_\theta^2 + \frac{1}{r^2 \sin^2\theta} p_\phi^2.$$

As an example of the first form given to the canonical transformation, where V depends on q and $\bar{q}$, we choose

$$V = \frac{c}{2} q^2 \cot \bar{q}.$$

Then we have

$$p = cq \cot \bar{q}$$

$$\bar{p} = \frac{c}{2} q^2 \frac{1}{\sin^2 \bar{q}}$$

or
$$q = \sqrt{\frac{2\bar{p}}{c}} \sin \bar{q}$$
$$p = \sqrt{2c\bar{p}} \cos \bar{q}.$$

Hence the expression
$$\tfrac{1}{2}(p^2 + c^2 q^2)$$
is transformed into $c\bar{p}$.

This example can be used to explain how the canonical transformations are employed in the integration of the equations of motion. For this we consider the *harmonic oscillator* in which

$$T = \frac{m}{2}\dot{q}^2 \qquad U = \frac{\kappa}{2}q^2.$$

Therefore
$$H = \frac{p^2}{2m} + \frac{\kappa}{2}q^2 = \frac{1}{2m}(p^2 + m\kappa q^2).$$

If in the last transformation given we place $c^2 = m\kappa$, H is transformed to $c\bar{p}/m$. This is the solution of the problem. For now $\bar{q} = \phi$ is a cyclic variable and we have

$$\bar{p} = \alpha$$
$$\bar{q} = \phi = \omega t + \beta, \qquad \omega = \frac{\partial H}{\partial \bar{p}} = \frac{c}{m} = \sqrt{\frac{\kappa}{m}}.$$

In the original coördinates the motion is represented by
$$q = \sqrt{\frac{2\alpha}{m\omega}} \sin(\omega t + \beta)$$
$$H = \omega\alpha.$$

LECTURE 3

The Hamilton-Jacobi partial differential equation — Action and angle variables — The quantum conditions.

In the same way we can now consider the most general case. Let us suppose that H does not depend explicitly on t. We shall denote constant momenta by α_k, the new variables which are linear functions of the time by ϕ_k, the number of degrees of freedom by f. Then we have to determine a function

$$S(q_1, q_2 \cdots q_f, \alpha_1, \alpha_2 \cdots \alpha_f)$$

so that, by the transformation,

$$\begin{cases} p_k = \dfrac{\partial}{\partial q_k} S(q_1, q_2 \cdots q_f, \alpha_1, \alpha_2 \cdots \alpha_f) \\ \phi_k = \dfrac{\partial}{\partial \alpha_k} S(q_1, q_2 \cdots q_f, \alpha_1, \alpha_2 \cdots \alpha_f) \end{cases} \qquad (1)$$

H becomes a function depending only on the α_k's,

$$W(\alpha_1, \alpha_2 \cdots \alpha_f).$$

Replacing p_k by its value in

$$H(q_1, q_2 \cdots p_1, p_2 \cdots)$$

we obtain the condition

$$H\!\left(q_1, q_2 \cdots q_f, \frac{\partial S}{\partial q_1}, \frac{\partial S}{\partial q_2} \cdots \frac{\partial S}{\partial q_f}\right) = W(\alpha_1, \alpha_2 \cdots \alpha_f). \qquad (2)$$

This expression can be looked upon as a partial differential equation for the determination of S. The problem is now to determine a so-called complete integral of this equation, that is, an integral which depends on $f - 1$ arbitrary constants $\alpha_2 \cdots \alpha_f$, where α_1 is to be identified with W, or, if no particular constant α_1 is to be privileged in this manner, then we must find an integral which depends on f constants $\alpha_1 \cdots \alpha_f$, among which there exists a relation

$$W = W(\alpha_1 \cdots \alpha_f).$$

The motion is then represented by

$$\phi_k = \omega_k t + \beta_k, \qquad \omega_k = \frac{\partial W}{\partial \alpha_k} \qquad (3)$$

We shall call Equation (2) the *Hamilton-Jacobi differential equation* and S the *action-function*. An important property of S is the following: We have

$$dS = \sum_k \frac{\partial S}{\partial q_k} dq_k = \Sigma p_k dq_k.$$

Therefore S is a line integral, taken along the orbit, from a fixed point Q_0 to a moving point Q.

$$S = \int_{Q_0}^{Q} \sum_k p_k dq_k. \qquad (4)$$

In Galilean-Newtonian mechanics this has a simple significance, because in this case,

$$2T = \sum_k p_k \dot{q}_k$$

and we have

$$S = 2 \int_{t_0}^{t} T\, dt = 2\overline{T}(t - t_0) \qquad (5)$$

where $\overline{T}$ is the time average of T.

We have seen already that the quantum theory is closely related to the periodic properties of the motion. In fact, Bohr's theory permits the definition of stationary states only for such motions as can be decomposed by harmonic analysis into periodic components. The astronomers call this class of motions "conditioned periodic." We prefer to call them "multiple periodic." These motions are defined in the following way: It is possible to introduce instead of variables q_k, p_k new variables w_k, I_k by means of the canonic transformation

$$p_k = \frac{\partial}{\partial q_k} S(q_1, I_1, q_2, I_2 \cdots q_f, I_f)$$

$$w_k = \frac{\partial}{\partial I_k} S(q_1, I_1, q_2, I_2, \cdots q_f, I_f)$$

which satisfy the following conditions:

THE STRUCTURE OF THE ATOM 23

(A) The position of the system depends periodically on w_k, with the fundamental period 1. That is, if the q_k's are uniquely determined by the position of the system, then they can be expanded in a Fourier series:

$$q_k = \sum_\tau C^{(k)}_\tau e^{2\pi i (w\tau)}$$

where τ represents a number of integers $\tau_1, \tau_2 \cdots \tau_f$ and we place

$$(w\tau) = w_1\tau_1 + w_2\tau_2 + \cdots + w_f\tau_f.$$

If one of the q_k's is an angle, it is not uniquely determined by the position of the system, but only within a multiple of a constant, as for instance 2π. Then the above condition of periodicity is also true except for a multiple of that constant.

(B) Hamilton's function can be transformed into a function W which depends only on the I_k's.

It follows that the I's are constants and the w's are linear functions of the time t,

$$w_k = \nu_k t + \beta_k.$$

The q's can therefore be represented by trigonometric series in t with the frequencies

$$\nu_1\tau_1 + \nu_2\tau_2 + \cdots + \nu_f\tau_f$$

where, according to the results obtained above,

$$\nu_k = \frac{\partial W}{\partial I_k}.$$

w_k, I_k are not yet uniquely determined by these conditions. For instance we can set

$$\overline{w}_k = w_k + f(I_1 \cdots I_f)$$

and

$$\overline{I}_k = I_k + C_k.$$

These form a canonic transformation, which is evidently compatible with the conditions (A) and (B). In order to exclude this indetermination we further set the condition:

(C) The function $\quad S^\times = S - \sum_k w_k I_k$

shall be periodic in w_k with the period 1:
$$S^\times = \sum_\tau C_\tau^\times e^{2\pi i(w\tau)}.$$

The canonic transformation in question can also be expressed by means of the function $S^\times$ as follows:
$$p_k = \frac{\partial}{\partial q_k} S^\times(q_1 \cdots q_f, w_1 \cdots w_f)$$
$$I_k = -\frac{\partial}{\partial w_k} S^\times(q_1 \cdots q_f, w_1 \cdots w_f).$$

Then indeed we can prove rigorously that w_k, I_k, which are called *angle* and *action variables*, are essentially uniquely determined by the conditions (A), (B) and (C). " Essentially " expresses the following: If we make a canonic transformation of the form
$$w_k = \sum_l c_{kl} \bar{w}_l$$
$$I_k = \sum_l c_{lk} \bar{I}_l$$

where the c's are whole numbers and the determinant $|c_{kl}| = \pm 1$, all the conditions (A), (B), (C) are still satisfied. Aside from this indetermination, however, w_k, I_k are really uniquely determined in all cases when the mechanical system is *not degenerate*, that is when there is no identical relation in ν_k of the form
$$\nu_1 \tau_1 + \nu_2 \tau_2 + \cdots + \nu_f \tau_f = 0$$
with the τ's whole numbers.

This theorem was first given by Burgers but his proof is not sufficient. A rigorous proof can be found in my book "Atommechanik"; this proof was given by my associate, F. Hund. This arbitrariness in the determination of the I_k's, whereby the latter are determined except for a whole-number transformation of determinant ± 1, is of essential importance for the applications of the quantum theory, for it is just these quantities that are equated to multiples of Planck's constant h; i.e.,
$$I_1 = n_1 h, \qquad I_2 = n_2 h, \cdots \qquad I_f = n_f h,$$
and from these equations it follows that also the $\bar{I}_k$'s are multiples of h.

LECTURE 4

Adiabatic invariants — The principle of correspondence.

In order to justify this method of quantization, it must be shown in the first place that the I's are adiabatic invariants. The general proof of this theorem was first outlined by Burgers and also by Krutkow; later more rigorous proofs were given by von Laue, Dirac and also by Jordan and myself. I shall not give here these rather complicated considerations, but shall only explain the significance of I and its adiabatic invariance using the example of the harmonic resonator. Using the Hamiltonian function,

$$H = \frac{1}{2\,m}(p^2 + m\kappa q^2),$$

and then applying a canonic transformation, we have found above a solution of the problem of the resonator, which, although not quite satisfying the conditions (A), (B), (C), is easy to transform into one which fulfils these conditions. It is only necessary to place

$$\phi = 2\,\pi w, \qquad \alpha = \frac{I}{2\,\pi}.$$

Then the transformation is

$$q = \sqrt{\frac{I}{\pi m \omega}}\sin 2\,\pi w, \qquad p = \sqrt{\frac{I m \omega}{\pi}}\cos 2\,\pi w$$

and the energy-function becomes

$$H = W = \omega \alpha = \frac{\omega}{2\,\pi} I = \nu I$$

where

$$\omega = 2\,\pi \nu$$

and also

$$w = \nu t + \delta, \qquad \nu = \frac{dW}{dI}.$$

As q is periodic in w with the period 1, and as H depends only on I, therefore the conditions (A) and (B) are fulfilled. To see whether condition (C) is also satisfied, we must only remember that the canonic transformation was found through the function

$$V = \frac{m\omega}{2} q^2 \cot 2\pi w$$

and then through the formulas,

$$p = \frac{\partial V}{\partial q}, \quad I = \frac{\partial V}{\partial w}.$$

This V is therefore identical with the $S^\times$ introduced above. It can be written in the form

$$V = S^\times = \frac{I}{2\pi m\omega} \sin 2\pi w \cos 2\pi w$$

and, since it is periodic, the condition (C) is also fulfilled.

The quantum condition

$$I = nh$$

gives therefore the energy levels,

$$W = nh\nu$$

in agreement with Planck's hypothesis. In order to verify that $I = W/\nu$ is really an adiabatic invariant, we represent the resonator by a pendulum swinging with small amplitude. Let m be the mass of the bob, l the length of the wire and g the acceleration of gravity. Suppose now that the length l is changed very slowly: the problem is to calculate how W and ν vary. The forces which stretch the wire for any value of the angle ϕ are the component of gravity $mg \cos\phi = mg(1 - \phi^2/2)$ and the centrifugal force $ml\dot\phi^2$. The work done in shortening the wire is therefore

$$A = -mg \int (1 - \phi^2/2) dl - ml \int \dot\phi^2 dl. \qquad (1)$$

Fig. 2

If the process of shortening is slow enough and has no period comparable with that of the pendulum, then it is permissible to introduce a mean amplitude, and we may write

$$dA = -mg\left(1 - \frac{\overline{\phi^2}}{2}\right)dl - ml\overline{\dot\phi^2}dl$$

where the dash denotes average over a period. The work done is now split up in two parts: $-mgdl$ is the work done in lifting the bob, and

$$dW = \left(\frac{mg}{2}\overline{\phi^2} - ml\overline{\dot\phi^2}\right)dl$$

is the increase in the energy of oscillation. Now we know that for harmonic oscillations:

$$\frac{W}{2} = \frac{m}{2}l^2\overline{\dot\phi^2} = \frac{m}{2}gl\overline{\phi^2}$$

and hence

$$dW = -\frac{W}{2l}dl.$$

Now, since ν is proportional to $1/\sqrt{l}$, therefore $\dfrac{d\nu}{\nu} = -\dfrac{dl}{2l}$ and

$$\frac{dW}{W} = \frac{d\nu}{\nu}$$

whence, by integration,

$$\frac{W}{\nu} = \text{constant} \tag{2}$$

which proves our theorem. The general proof of adiabatic invariance consists essentially of quite analogous considerations.

As another important example let us consider the rotator, that is, a body which can be rotated about an axis. If A is the moment of inertia with respect to the axis and ϕ the angle of rotation, then we have

$$H = \frac{A}{2}\dot\phi^2$$

whence it follows, for the momentum p corresponding to ϕ

$$p = A\dot\phi.$$

p is the angular momentum and we have

$$H = \frac{p^2}{2A}.$$

ϕ is therefore a cyclic variable and

$$p = \text{constant}.$$

If we set $\phi = 2\pi w$, the position of the system is a periodic function of w of period 1. The canonic transformation

$$(\phi, p) \rightarrow (w, I)$$

is evidently characterized by the function $S = \phi I/2\pi$ and has the form

$$p = \frac{\partial S}{\partial \phi} = \frac{I}{2\pi}, \qquad w = \frac{\partial S}{\partial I} = \frac{\phi}{2\pi},$$

whence it follows that $S^\times = S - wI = 0$ is a periodic function. Finally, we obtain

$$H = W = \frac{I^2}{8\pi^2 A}.$$

The conditions (A), (B), (C) are fulfilled and we have to set

$$I = h\nu,$$

which gives the energy levels

$$W = \frac{h^2}{8\pi^2 A} n^2. \tag{3}$$

This model is applied to the explanation of the band spectra of molecules. If a molecule rotates about a fixed axis, the emitted frequencies, according to Bohr, are given by the relation $\nu = \frac{1}{h}(W_m - W_n) = \frac{h}{8\pi^2 A}(m^2 - n^2)$, but, as in the case of the oscillator, the number of different frequencies given by this formula is too large. Out of these frequencies we must choose certain ones by the Principle of Correspondence. For this purpose we consider a component of the electric moment of the rotator. Evidently, in this case the motion is also given by a simple harmonic oscillation, whence we conclude, as above, that there are no other jumps of n than those where n changes by ± 1, i.e., that $n - m = \pm 1$. Introducing this restriction, we obtain for the emitted frequencies (placing $m - n = 1$, or $m = n + 1$):

$$\nu = \frac{h}{8\pi^2 A}((n+1)^2 - n^2) = \frac{h}{8\pi^2 A}(2n+1)$$

$$\nu = \frac{h}{4\pi^2 A}(n + \tfrac{1}{2}).$$

The rotation frequency of the rotator itself is given by

$$\nu_0 = \frac{dW}{dI} = \frac{I}{4\pi^2 A} = \frac{nh}{4\pi^2 A}.$$

Therefore, as n increases, the relative difference between the rotation and the emitted frequency becomes smaller. In both cases we have an equidistant series of frequencies and, indeed, the band spectrum emitted by a rotating molecule appears as a first approximation to consist of such a series.

We shall not go further into this problem, but instead will now consider the general relation which exists, according to the Principle of Correspondence, between the frequencies and the intensities of the spectral lines calculated classically and the corresponding quantities calculated according to the quantum theory. We consider the electric moment of the system having a Fourier expansion analogous to that of the coördinates

$$\mathbf{M} = \sum_k e_k \mathbf{r}_k = \sum_\tau \mathbf{C}_\tau e^{2\pi i(w\tau)} = \sum_\tau \mathbf{C}_\tau e^{2\pi i[(\nu\tau)t + (\delta_\tau)]} \tag{4}$$

The frequencies can be written

$$\nu_{cl} = (\nu\tau) = \sum_k \nu_k \tau_k = \sum_k \tau_k \frac{\partial W}{\partial I_k}. \tag{5}$$

Let a stationary state be determined by

$$I_k^{(1)} = n_k^{(1)} h$$

and another by

$$I_k^{(2)} = n_k^{(2)} h;$$

then we can consider in the I_k-space of f dimensions the two points connected by the straight line

$$I_k = I_k^{(1)} + \tau_k \lambda; \quad 0 \leq \lambda \leq h$$

where $\quad \tau_k = n_k^{(2)} - n_k^{(1)}$.

Then

$$\frac{dI_k}{d\lambda} = \tau_k$$

and

$$\nu_{cl} = \sum_k \frac{\partial W}{\partial I_k} \frac{dI_k}{dt} = \frac{dW}{d\lambda}.$$

On the other hand the quantum frequency is

$$\nu_{qu} = \frac{W_1 - W_2}{h} \qquad (6)$$

and the relation between the frequencies in the classical and in the quantum theory is the same as that between derivative and difference-ratio. It is also possible to consider the quantum-theory frequencies as the straight-line mean of the classical frequencies, as follows,

$$\nu_{qu} = \frac{1}{h} \int dW = \frac{1}{h} \int_0^h \frac{dW}{d\lambda} d\lambda = \frac{1}{h} \int_0^h \nu_{cl} d\lambda. \qquad (7)$$

If the changes of the quantum numbers are small compared with the numbers themselves, the two expressions for ν_{qu} and ν_{cl} respectively differ very little. As to the intensities, we expect that they vary approximately in the same way as the quantities $|\mathbf{C}_\tau|^2$, where $\mathbf{C}_\tau$ is a function of the I_k's and of $\tau_k = n_k^{(1)} - n_k^{(2)}$ $= (I_k^{(1)} - I_k^{(2)})/h$. It is seen that this statement has a definite meaning only if n_k is large, because only in this case is it immaterial whether we place in $\mathbf{C}_\tau(I) = \mathbf{C}_{n^{(1)} - n^{(2)}}(n)$ for n the initial value $n^{(1)}$ or the final value $n^{(2)}$. On the other hand, this statement has a unique meaning if $\mathbf{C}_\tau(I)$ is identically zero for all I's, for then we expect that a jump of τ does not occur. In other cases the difficulty has been evaded by taking a suitable mean of $\mathbf{C}_\tau(I)$ over the values of I between the initial and the final states. By this method, Kramers has succeeded in representing satisfactorily the results of observations in certain

THE STRUCTURE OF THE ATOM 31

cases. It is not satisfactory in principle that we should not find in the quantum theory, in the form here presented, a unique determination of the intensities. This is one of the main reasons which led us to formulate our new quantum theory, where this difficulty is overcome.

LECTURE 5

Degenerate systems — Secular perturbations — The quantum integrals.

We now say a few words about the case, left aside so far, of degeneration, that is, that in which there exist identical relations in the I_k's of the form

$$(\nu\tau) = \nu_1\tau_1 + \nu_2\tau_2 + \cdots + \nu_n\tau_n = 0. \quad (1)$$

Then our theorem of uniqueness no longer holds, and it is no longer possible to formulate the quantum conditions in the form

$$I_k = n_k h.$$

Such is the case, for instance, in the harmonic oscillator of two degrees of freedom.

$$H = \frac{1}{2m}(p_x^2 + p_y^2) + \frac{m}{2}(\omega_x^2 x^2 + \omega_y^2 y^2).$$

The solution of the equations of motion can be written down immediately, because the two coördinates are separable. We obtain

$$x = \sqrt{\frac{I_x}{\pi\omega_x m}} \sin 2\pi w_x, \qquad p_x = \sqrt{\frac{\omega_x m I_x}{\pi}} \cos 2\pi w_x$$

$$y = \sqrt{\frac{I_y}{\pi\omega_y m}} \sin 2\pi w_y, \qquad p_y = \sqrt{\frac{\omega_y m I_y}{\pi}} \cos 2\pi w_y.$$

Here I_x, w_x; I_y, w_y are two conjugate pairs of action and angle variables. If now w_x and w_y are *not* commensurable the motion given by placing (Fig. 3)

$$w_x = \omega_x t + \delta_x, \qquad w_y = \omega_y t + \delta_y$$

is a so-called Lissajous figure, in which the path comes as near as desired to any point within a rectangle. But if a relation of the form

$$\tau_x \omega_x + \tau_y \omega_y = 0$$

THE STRUCTURE OF THE ATOM 33

exists, for instance if (Fig. 4.)

$$\omega_x = \omega_y = \omega_0 = 2\pi\nu_0,$$

then the orbit is simple periodic (ellipse). We can now rotate the system of coördinates arbitrarily without changing the form of the solution. But in doing so the sides of the rectangle change continuously and so do the magnitudes $\sqrt{I_x}$, $\sqrt{I_y}$,

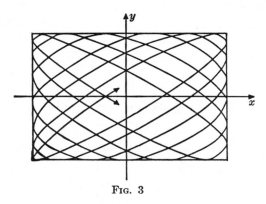

Fig. 3

which differ from them only by the constant factor $1/\sqrt{\pi\omega m}$. It is therefore impossible to place I_x and I_y proportional to whole numbers n_x, n_y. The diagonal of the rectangle, however, that is the square root of the quantity

$$(\sqrt{I_x})^2 + (\sqrt{I_y})^2 = I_x + I_y = I,$$

remains invariant for such a rotation. We can therefore set

$$I_x + I_y = nh$$

whereby the total energy

$$W = \frac{\omega_0}{2\pi}(I_x + I_y) = \nu_0 I$$

is uniquely determined. $W = nh\nu_0$ has therefore exactly the same value as for the linear oscillator. We can describe this behavior as follows: If we introduce, instead of I_x and I_y, two new variables, $I_x + I_y = I$ and $I_x - I_y = I'$, then two new conjugate angle variables, w, w', correspond to the latter, with the frequencies,

$$\nu = \frac{dW}{dI} = \nu_0, \qquad \nu' = \frac{dW}{dI'} = 0,$$

and we can only quantize the variable I which alone appears in W and which therefore alone corresponds to a frequency different from zero.

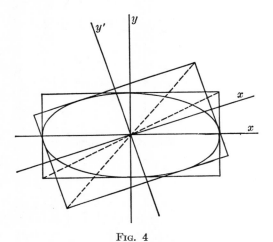

Fig. 4

The following rule holds in general: In cases of degeneration it is always possible to attain, by means of a linear whole-number transformation of determinant ± 1, that $H = W$ shall depend only on a number s of I-variables, among which there are no commensurability relations. We call such variables I_α. To these variables correspond s frequencies ν_α different from zero, while the other $f - s$ frequencies ν_ρ vanish. Only such variables I_α are to be equated to multiples of h. Bohr calls s the *degree of periodicity* of the system.

Evidently we can increase the degree of periodicity of a system by introducing perturbing forces, for instance by placing the system in an electric or in a magnetic field. Then the original energy-function, which we shall call H_0, is increased by an additional energy which we shall call "perturbation energy" and denote by λH_1, where λ is a measure of the magnitude of the additional energy. If the perturbation is small, that is, if λ is small, then there is a simple process whereby the new motions which are to be added to the system which was originally degenerate can be calculated. The influence of the perturbation energy is to change slightly all the magnitudes w, I, but the influence is different for these two kinds of variables. Those angle variables w_ρ which belong to the zero frequencies of the unperturbed system, and which therefore were constant, now change slowly with frequencies which are proportional to λ. The other angular variables w_α will only undergo small variations of their frequencies. If we take the w^0, I^0's of the unperturbed system as initial variables for the perturbation problem, then we have

$$H = H_0(I_\alpha^0) + \lambda H_1(I_\alpha^0, w_\alpha^0; I_\rho^0, w_\rho^0) \qquad (2)$$

and w_α^0 will complete its period many times during a period of w_ρ^0. Therefore an approximation can be made by taking an average over w_α^0

$$\overline{H} = H_0(I_\alpha^0) + \lambda \overline{H}_1(I_\alpha^0; I_\rho^0, w_\rho^0). \qquad (3)$$

This function can be considered as the energy-function of a new problem of motion for the formerly degenerate variables I_ρ^0, w_ρ^0. It is required to solve the equations of motion

$$\dot{w}_\rho^0 = \lambda \frac{\partial \overline{H}_1}{\partial I_\rho^0}, \qquad \dot{I}_\rho^0 = -\lambda \frac{\partial \overline{H}_1}{\partial w_\rho^0}, \qquad (4)$$

that is, to find a canonical substitution

$$(I_\rho^0, w_\rho^0) \rightarrow (I_\rho, w_\rho)$$

such that $\overline{H}_1$ is transformed into a function W_1 which depends only on I, i.e., (I_α is written for I_α^0).

$$H = W_0(I_\alpha) + \lambda W_1(I_\alpha, I_{\rho'}).$$

The perturbation frequencies are then

$$\nu_\alpha = \frac{\partial W_0}{\partial I_\alpha} + \lambda \frac{\partial W_1}{\partial I_\alpha}, \qquad \nu_\rho = \lambda \frac{\partial W_1}{\partial I_\rho}.$$

36 PROBLEMS OF ATOMIC DYNAMICS

The name "secular perturbations" is given in celestial mechanics to the slow motions of frequencies ν_ρ. We shall only touch upon the question of how the angle and action variables can be actually found in given cases. The method of separation of variables is often used: It is applicable if it is possible to find canonic variables p_k, q_k for which the Hamilton-Jacobi differential equation can be solved by setting

$$S = S_1(q_1) + S_2(q_2) + \cdots + S_f(q_f). \qquad (5)$$

Then

$$p_k = \frac{\partial S_k}{\partial q_k} \qquad (6)$$

is a function of q_k only and we can show that the integrals taken over a period

$$I_k = \int_o p_k dq_k = \int_o \frac{\partial S_k}{\partial q_k} dq_k \qquad (7)$$

are the action variables. The functions $S_k(q_k)$ depend also on these constants I_k. Applying a canonic transformation, in which the I_k's enter as new action variables, the corresponding new angle variables are defined by

$$w_k = \frac{\partial S}{\partial I_k} = \sum_l \frac{\partial S_l}{\partial I_k}. \qquad (8)$$

Since S satisfies the Hamilton-Jacobi differential equation, H is transformed into a function $W(I_1 \cdots I_f)$ and the condition (A) is satisfied.

If any coördinate q_h varies once between its limits, while the other coördinates q_k are kept constant, the change of any variable w_k is

$$\Delta_h w_k = \int_o \frac{\partial w_k}{\partial q_h} dq_h.$$

Now,

$$\frac{\partial w_k}{\partial q_h} = \sum_l \frac{\partial^2 S_l}{\partial I_k \partial q_h} = \frac{\partial}{\partial I_k} \sum_l \frac{\partial S_l}{\partial q_h} = \frac{\partial}{\partial I_k} \frac{\partial S_h}{\partial q_h}$$

THE STRUCTURE OF THE ATOM 37

whence

$$\Delta_h w_k = \frac{\partial}{\partial I_k} \int_0 \frac{\partial S_h}{\partial q_h} dq_h = \frac{\partial I_h}{\partial I_k} = \begin{cases} 1 \text{ for } h = k. \\ 0 \text{ for } h \neq k. \end{cases} \quad (9)$$

If any point $q_1^0 \cdots q_f^0$ in q-space, to which corresponds the point $w_1^0 \cdots w_f^0$ in w-space, describes a closed curve, then the point w need not return to its original position, but the end point is given by an expression of the form $w_k^0 + (\tau_1 w_1^0 + \cdots + \tau_f w_f^0)$ where the τ's are whole numbers. The q's are therefore periodic functions of the w's, with the fundamental period 1. Condition (B) is thus satisfied.

According to the definition of I_k, S increases by the amount I_k every time that q_k varies over a cycle, the other variables being held constant. As w_k increases by 1 at the same time, the function $S^\times = S - \sum_k w_k I_k$ remains unchanged. Therefore, it is periodic, and the condition (C) is satisfied, whence it is proved that w, I are the angle and action variables.

Many authors introduce the quanta by this integral definition, but it appears to me, as to Bohr, better to define them generally by the properties of periodicity, that is, by the three conditions (A), (B), (C).

LECTURE 6

Bohr's theory of the hydrogen atom — Relativity effect and fine structure — Stark and Zeeman effects.

After these general considerations we now take up the applications to the theory of atomic structure. As you know it was with the *hydrogen atom* that Bohr first developed his ideas. We have in this case one nucleus and one electron, that is, a two-body problem which can be reduced, as you know, to a one-body problem: the motion of a point around a fixed center of attraction. If r, ϕ, and θ are the polar coördinates of the electron relative to the nucleus, and if we place

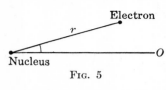

Fig. 5

$$\frac{1}{\mu} = \frac{1}{M} + \frac{1}{m}$$

where M is the mass of the nucleus, and m that of the electron, we have

$$H = \frac{\mu}{2}(\dot{r}^2 + r^2\dot{\theta}^2 + r^2\dot{\phi}^2 \sin^2 \theta) + U(r).$$

The potential energy of the Coulomb force between a nucleus carrying a Z-fold charge and an electron is

$$U(r) = -\frac{Ze^2}{r}$$

but we shall also consider general central forces with an arbitrary function $U(r)$.

Introducing the momenta we obtain

$$H = \frac{1}{2\mu}\left(p_r^2 + \frac{1}{r^2}p_\theta^2 + \frac{p_\phi^2}{r^2 \sin^2 \theta}\right) + U(r). \qquad (1)$$

The corresponding Hamilton-Jacobi differential equation can be easily solved by separation of variables. In the case of

THE STRUCTURE OF THE ATOM

Coulomb's law we obtain the well-known Keplerian motions; of these only periodic orbits, i.e., ellipses, come into consideration in the quantum theory. It is seen at once that the motion is doubly degenerate for it has three degrees of freedom but is only simple-periodic. There is therefore only one action quantity I and one quantum condition. Calculation shows that I is related to the major axis a of the ellipse by the formula

$$a = \frac{I^2}{4\pi^2 \mu e^2 Z}$$

and for the energy we obtain

$$W = -\frac{2\pi^2 \mu e^4 Z^2}{I^2}. \tag{2}$$

Referred to a system of axes directed along the axes of the ellipse, the motion is represented by simple Fourier series

$$\frac{x}{a} = -\frac{3}{2}\epsilon + \sum_{\tau=1}^{\infty} C_\tau(\epsilon) \cos 2\pi w\tau$$
$$\frac{y}{a} = \sum_{\tau=1}^{\infty} D_\tau(\epsilon) \sin 2\pi w\tau \tag{3}$$

the coefficients of which are continuous functions of the eccentricity ϵ. The angle variable w is, except for the factor 2π, the "mean anomaly" of astronomers.

These were the starting formulas for Bohr's theory of the hydrogen atom. Placing

$$I = nh$$

and

$$R = \frac{2\pi^2 \mu e^4}{h^3} \tag{4}$$

he found

$$W = -\frac{RhZ^2}{n^2} \tag{5}$$

and obtained for the frequencies of the emitted light

$$\nu = \frac{1}{h}(W_1 - W_2) = RZ^2\left(\frac{1}{n_2^2} - \frac{1}{n_1^2}\right). \tag{6}$$

For the hydrogen atom $Z = 1$ and this formula gives in fact all the known lines of hydrogen, in particular the Balmer series ($n_2 = 2$),

$$\nu = R_H\left(\frac{1}{4} - \frac{1}{n_1^2}\right) \qquad n_1 = 3, 4, 5 \cdots.$$

The formula gives not only the dependence on n_1 but, what is more important, the correct value of R_H. To calculate the latter we have to replace μ by the expression

$$\mu = \frac{mM}{m+M} = m\frac{1}{1+\frac{m}{M}}.$$

We may therefore write

$$R_H = R_\infty \frac{1}{1+\frac{m}{M}}, \qquad R_\infty = \frac{2\pi^2 me^4}{h^3} = 3.28 \times 10^{15} \text{ sec.}^{-1}$$

in which e, m and h are replaced by the best experimental values. Neglecting the small fraction m/M, which is about 1/1830, we obtain, dividing by the velocity of light, $c = 3 \times 10^{10}$ cm./sec.,

$$\frac{R_H}{c} = \frac{3.28 \times 10^{15}}{c} = 1.09 \times 10^5 \text{ cm.}^{-1}$$

while spectroscopic measurements give 109678 cm.$^{-1}$.

The series given by $n_2 = 1$, $n_2 = 2$, $n_2 = 3$, $n_2 = 4$, $n_2 = 5$, have also been measured (by Lyman, Paschen, Brackett). Moreover Bohr was justified in maintaining that the series which is obtained by putting $Z = 2$, and which had until then been ascribed to hydrogen, must belong to ionized helium,

$$\nu = 4 R_{He}\left(\frac{1}{n_2^2} - \frac{1}{n_1^2}\right) = R_{He}\left(\frac{1}{\left(\frac{n_2}{2}\right)^2} - \frac{1}{\left(\frac{n_1}{2}\right)^2}\right).$$

The fraction m/M is now four times smaller than for the H-atom, because the He-atom is four times heavier. Therefore, the lines for same n_1 and n_2 do not exactly coincide with the

THE STRUCTURE OF THE ATOM 41

hydrogen lines. This separation is observed experimentally and now we are certain that the spectrum is that of ionized helium, to be sure the most beautiful result of Bohr's theory.

Bohr's theory of all other spectra may be briefly described as an attempt to consider them as modifications of the hydrogen spectrum. Two lines of attack are to be distinguished here; the first is to calculate the influence of secondary effects on the hydrogen atom: The dependence of mass on velocity is taken into consideration and gives the fine structure of the lines, then the influence of external electric and magnetic fields (Stark and Zeeman effects). The second line of attack leads to the study of other atoms and, together with it, to a theoretical systematic study of relations among the atoms and of the periodic system of the elements. Let me speak about the first line of attack.

Sommerfeld was the first to point out and carry through the idea that the variation of mass demanded by the *theory of relativity* must have an effect on the spectrum. He replaced the classical energy-function by the relativistic one given in the first lecture:

$$H = m_0 c^2 \left[\sqrt{1 + \frac{p^2}{m_0 c^2}} - 1 \right] - \frac{e^2 Z}{r}. \quad (7)$$

On account of the smallness of the effect it is sufficient to take into account the first term in the expansion in powers of $\frac{p^2}{m_0 c^2}$ and write:

$$H = H_0 + H_1$$

where H_0 is the classical energy-function and

$$H_1 = -\frac{1}{8 \, m_0^3 c^2} (p_x^2 + p_y^2 + p_z^2)^2$$

is the perturbation function.

The law of areas holds also in relativistic mechanics. Therefore the orbit is plane, but the plane orbit is no longer a simple periodic ellipse, but is transformed into a "rose-shaped" figure. The motion can also be described as an elliptic motion the major axis of which is rotating uniformly. The law of precession of the perihelion is found, following the method of secular perturbations, by taking the average of the function H_1 over the unperturbed motion

$$\overline{H}_1 = -\frac{RhZ^2}{n^2}\frac{\alpha^2 Z^2}{n^2}\left(\frac{I}{I'} - \frac{3}{4}\right) \quad (8)$$

where

$$\alpha = \frac{2\pi e^2}{hc} = 7.29 \times 10^{-3}$$

is a numerical constant, I' is the angle variable conjugate to the azimuth w' of the major axis and depends on the eccentricity ϵ by the simple relation

$$I' = I\sqrt{1 - \epsilon^2}.$$

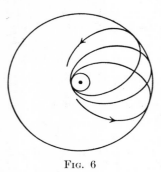

FIG. 6

Since w' does not appear in H_1, therefore it is a cyclic variable and we have the new quantum condition

$$I' = kh. \quad (9)$$

k is called the azimuthal quantum number to distinguish it from the main quantum number n. k is always less than or equal to n. The total energy becomes

$$W = -\frac{RhZ^2}{n^2}\left[1 + \frac{\alpha^2 Z^2}{n^2}\left(\frac{n}{k} - \frac{3}{4}\right)\right]. \quad (10)$$

This formula expresses that every term of the unperturbed spectrum is separated into a number of terms which correspond to the values $k = 1, 2 \cdots n$. From this arises a splitting of the spectral lines,

$$\nu = \frac{1}{h}[W(n_1, k_1) - W(n_2, k_2)]$$

in such a way that k is changed only by ± 1, for the rotation of the perihelion determined by $I' = kh$ is a simple harmonic motion. This split of spectral lines was predicted by Som-

THE STRUCTURE OF THE ATOM 43

merfeld and experimentally verified for hydrogen and ionized helium, not only for the number of lines but also for the absolute value of the separation.* Kramers has also calculated the intensity of the lines by means of the principle of correspondence and found good agreement with observations.

The influence of an external electric field, i.e., the *Stark effect*, can be treated in quite an analogous way. The perturbation energy is in this case,

$$H_1 = eEz \qquad (11)$$

where z is the coördinate of the electron along the z-axis taken parallel to the field E. It is therefore simply required to calculate the average of z. This depends not only on the position of the major axis of the ellipse in the orbital plane, but also on the orientation of this plane in space. It can be shown, however, that the problem of the secular perturbation can be reduced to one of one degree of freedom. The calculation gives for the energy,

$$W = -\frac{RhZ^2}{n^2} \pm \frac{3\,Eh^2}{8\,\pi^2\mu eZ}\,nn_e \qquad (12)$$

where n_e is a new quantum number which varies between $-(n-1)$ and $(n-1)$. The motion itself can be described in the following way. If we calculate the electric "center of gravity" S of the electron in its orbit, that is, the average value of its coördinates over one revolution, we find that it is on the major axis at a distance $\frac{3}{2}\,a\epsilon$ from the nucleus O in the direction towards the aphelion. On account of secular

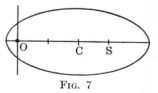

Fig. 7

perturbations, this point moves with simple harmonic motion in a plane perpendicular to the field E, whence it follows that n_e can only change by ± 1. The split of spectral lines is completely determined hereby, in good agreement with the results of experiment, also as regards the intensities calculated by Kramers.

* For hydrogen the quantitative results are not yet quite certain.

44 PROBLEMS OF ATOMIC DYNAMICS

For the *Zeeman effect* the calculation is still simpler and moreover can be carried out for atoms with an arbitrary number of electrons. The expression given earlier for the energy in a magnetic field is, neglecting terms containing the square of the field strength (Equation (14), Lecture 2):

$$H = H_0 + \frac{e}{c\mu} \Sigma \mathbf{A} \cdot \mathbf{p} \qquad (13)$$

where H_0 is the energy of the unperturbed system. The vector potential of a homogeneous field is

$$\mathbf{A} = \tfrac{1}{2}\,\mathbf{H}\times\mathbf{r}.$$

Therefore

$$\Sigma \mathbf{A} \cdot \mathbf{p} = \tfrac{1}{2}\,\Sigma \mathbf{H}\times\mathbf{r} \cdot \mathbf{p} = \tfrac{1}{2}\,\mathbf{H}\Sigma\mathbf{r}\times\mathbf{p} = \tfrac{1}{2}\,|\mathbf{H}|P_\phi$$

where P_ϕ is the component of angular momentum $\mathbf{P} = \Sigma \mathbf{r}\times\mathbf{p}$ parallel to the field. For the unperturbed system the angular momentum is constant in magnitude $|\mathbf{P}|$ and direction. It is easy to see that $2\,\pi|\mathbf{P}|$ is an action integral. We place therefore

$$2\,\pi|\mathbf{P}| = jh. \qquad (14)$$

The components of $\mathbf{P}$ are also constant, but they are evidently conjugate to degenerate angle variables. In the magnetic field the degeneration of the angle ϕ, which fixes the position of the plane determined by the field and the angular momentum with respect to a fixed plane parallel to the direction of the field, is removed and the system precesses around the direction of the field. P_ϕ is conjugate to ϕ, as easily seen. We have therefore the new quantum condition

$$2\,\pi P_\phi = mh. \qquad (15)$$

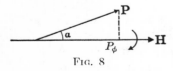

Fig. 8

If α is the angle between the angular momentum and the direction of the field, then evidently

$$\cos\alpha = \frac{P_\phi}{|\mathbf{P}|} = \frac{m}{j}.$$

The axis of angular momentum can therefore only take $2j + 1$ different directions ($m = -j, \cdots +j$) with respect to the

THE STRUCTURE OF THE ATOM

direction of the field axis. We shall call this result, following Sommerfeld, "directional quantization."

The energy is

$$H = W_0 \pm \frac{eh}{4\,\pi\mu c}\,|\mathbf{H}|\,m \tag{16}$$

whence the number of revolutions of the axis of angular momentum, the so-called "Larmor frequency," is

$$\nu_m = \frac{\partial H}{\partial\,2\,\pi P_\phi} = \frac{1}{h}\frac{\partial H}{\partial m} = \frac{e\,|\mathbf{H}|}{4\,\pi\mu c} = 4.70 \times 10^{-5}\,|\mathbf{H}|\ \text{cm.}^{-1}. \tag{17}$$

Precession does not influence the components of the motion of the electrons in the direction parallel to the field. Therefore there is no additional term in the z-component of the electric moment and light oscillating parallel to z corresponds to jumps for which m does not change. The components of motion perpendicular to the field are altered however by simple rotations in one direction or the other, hence the emitted light must be decomposed into two circularly polarized waves in opposite directions, to which correspond the jumps

$$m \to m \pm 1.$$

We obtain therefore the classical Zeeman triplet without any change. This contradicts experiment, however, for in most cases spectral lines are split up in a much more complicated way. Bohr's theory in its present form gives no explanation of this more complicated effect. According to it we should expect in all cases and for every atom normal Larmor precession and the normal spectral triplet. At this point many attempts have been made to change the theory. Starting from Sommerfeld's researches, Landé has succeeded in decomposing the observed Zeeman separation of most spectral lines into terms and discovered their relation to the periodic system of the elements. Heisenberg, Pauli and many others have investigated this problem further. The essential result of all these investigations is that the so-called "abnormal" Zeeman effect — which is, however, certainly the normal case — finds no place in the semi-classic theory which we have developed here.

The positive result is that the Zeeman effect is closely connected with the construction of atoms out of electrons describing orbits to which correspond fixed quantum numbers. We shall now treat this problem of the arrangement of electronic orbits in the atom, following the method of Bohr, who considered the series spectra as modifications of the hydrogen spectrum.

LECTURE 7

Attempts towards a theory of the helium atom and reasons for their failure — Bohr's semi-empirical theory of the structure of higher atoms — The optical electron and the Rydberg-Ritz formula for spectral series — The classification of series — The main quantum numbers of the alkali atoms in the unexcited state.

The most obvious way of finding an exact theory of atomic structure would be to consider successively the simplest atoms, helium, lithium, etc., following hydrogen in the series of the elements. This has been tried, but even the first step from the hydrogen to the helium atom proved unsuccessful. The helium atom is an instance of the three-body problem: one nucleus and two electrons. It is well known that the three-body problem has greatly perplexed astronomers and that it has not been possible to represent the motion by analytical expressions (series) which really permit a survey of the motion at all times. In the case of atomic structure, conditions are even less favorable, for in celestial mechanics there is at least the advantage that the attraction towards the central body is much greater than the other attractions on account of the preponderant mass of the sun, so that all these other attractions can be looked upon as small "perturbations." In atomic mechanics, however, all the attractions and repulsions of electric charges are of the same order of magnitude. On the other hand, the atomic problem has an advantage of a different kind, precisely on account of the postulate of the quantum theory, that only certain "stationary" orbits come into consideration. It has been shown that the quantum conditions allow only very simple types of orbits, because they exclude certain librations (oscillations).

Based on this result, attempts have been made to find the stationary orbits for the helium atom, and calculate its energy levels. The line of attack has been along two directions: Some investigators have considered the normal state of the helium atom (Bohr, Kramers, van Vleck), others the excited state, in which one electron is in the nearest orbit to the nucleus and the other revolves in a very distant orbit (van Vleck, Born and Heisenberg). Both calculations give incorrect results: The calculated energy of the normal state does not agree with experimental results (ionization energy of the normal helium atom), and the calculated term system for the excited states is different from that observed, qualitatively as well as quantitatively.

After all, no other result could be expected, for the validity of the frequency condition is sufficient to show conclusively that in the realm of atomic processes the laws of classical theories (geometry, kinematics or mechanics, electrodynamics) are not right. That in certain simple cases, as for a single electron, they give partially correct results is, in fact, more astonishing than that they fail in the more complicated cases of several electrons. This failure of the theory in the case of interactions among several electrons is evidently connected with the following fact: We know that electrons react quite unclassically to light waves, because the latter produce quantum jumps. In a system made up of several electrons, each electron is in the oscillating field due to all other electrons and the periods of these fields are of the same order of magnitude as those of light waves, therefore we have no reason to expect that the electron should react classically to this oscillating field. This point of view gives grounds for understanding why we obtain, by the classical theory, correct results in many cases of the one-electron problem.

Appreciating these difficulties, Bohr has given up the attempt to construct a truly deductive theory, and, instead, has endeavored, with the greatest success, to discover, by interpretation of facts, above all of the facts relating to the spectra and the chemical and magnetic properties of the atoms, something about the arrangement of the electrons. The starting point was the observation of the fact that the spectra of certain atoms are of a type quite similar to the hydrogen spectrum. The lines, or better, the terms form series quite similar to the series of terms $\frac{R}{n^2}$ of the H-atom. Rydberg, for instance,

showed that in many cases expressions of the form $\dfrac{R}{(n + \delta)^2}$
with δ = constant, are sufficient for expressing the terms. This is the case for the alkali metals, for part of the lines of Cu, Ag, Au, and for other similar cases, all of which share chemical properties which indicate the easy detachment of one electron. From this Bohr concludes that all these spectra, as that of hydrogen, are produced by the jumps of *one* electron, the " optical electron." This electron, however, does not move around a simple nucleus, but around a *core* consisting of the nucleus and all the remaining electrons. If the optical electron is more and more strongly excited — that is, brought up to levels of higher energy — the state of total separation called ionization is gradually reached and then the core is left as an "ion." This argument agrees with the results of the chemists, as formulated by Lewis, Langmuir and Kossel. According to these theories the ions of the alkali metals have the same structure as the atoms of the neighboring inert gases: The latter are the most stable closed electron configurations.

Now it can be shown that the orbit of the optical electron in the lower stationary states must penetrate into the core, for otherwise the terms would differ but little from those of the H-atom. Moreover, the radii of the ions are fairly well known from the theory of electrolytes and of polar crystals and it can be estimated, by a method to be given shortly, that the orbit of the optical electron must go through the core (Schrödinger, Bohr).

In postulating such "penetrating orbits," a step is taken which is incompatible with ordinary mechanics, for following our quantization rules we must assume that the orbit of the optical electron is exactly periodic, but this cannot be understood from the standpoint of mechanics because of the intensive interaction with the inner electrons. It would be necessary to assume that the whole electronic structure is rigorously periodic in every quantized state and it is very questionable whether there exist such solutions of the mechanical equations. If in spite of this we wish to describe the paths of the optical electron within the realm of our theory and more or less approximately, this may be done following Bohr and in a purely formal way, by replacing the action of the core on the electron by a central force and neglecting altogether the reaction of the electron on the atomic core. Then the conservation of energy is always satisfied for the electron alone and we have to do, as hitherto, with a one-body problem. Conservation of angular momentum also holds and the orbit is plane.

We can show, following Bohr, that the terms must be approximately expressible by formulas of Rydberg type, or more accurately of Ritz type, assuming that the core is small compared with the size of the orbit of the optical electron. The outer part of the orbit differs only slightly from a Keplerian ellipse, while the inner part is an oval of small radius of curvature, because there the electron comes in a region of strong nuclear attraction (Fig. 9).

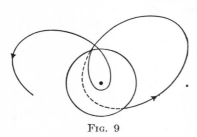

Fig. 9

If we replace the outer part of the orbit by an ellipse, then its energy becomes

$$W = -\frac{RhZ^{\times 2}}{n^{\times 2}}. \tag{1}$$

In this formula $n^\times$ depends on the aphelion distance $2\,a^\times$, in the same way as was given above for other quantized orbits, i.e.

$$a^\times = \frac{h^2}{4\,\pi^2\mu e^2 Z^\times}\, n^{\times 2} \tag{2}$$

and $Z^\times$ is the "effective nuclear charge," that is the difference between the charge of the nucleus and of the screening electrons of the ion. $n^\times$ need not be an integer, for it is not the main quantum number of the whole orbit; if we call the latter n then the frequency of the motion from one aphelion to the next is given by

$$\nu = \frac{\partial W}{\partial I} = \frac{1}{h}\frac{\partial W}{\partial n}. \tag{3}$$

THE STRUCTURE OF THE ATOM 51

On account of our assumption that the core is small, the time of revolution $\frac{1}{\nu}$ of the whole orbit differs slightly from the time of revolution $\frac{1}{\nu^\times}$ of the ellipse replacing the actual path. The latter is given by

$$\nu^\times = \frac{1}{h}\frac{\partial W}{\partial n^\times} = \frac{2\,RZ^{\times 2}}{n^{\times 3}}. \tag{4}$$

We therefore place

$$\frac{1}{\nu} = \frac{1}{\nu^\times} + b \quad \text{or} \quad \frac{\nu^\times}{\nu} = 1 + b\nu^\times = 1 + \frac{2\,RbZ^{\times 2}}{n^{\times 3}}$$

and consider b to be approximately constant; then we have

$$\frac{dn}{dn^\times} = \frac{\nu^\times}{\nu} = 1 + \frac{2\,RbZ^\times}{n^{\times 3}}$$

which integrated gives

$$n = n^\times - \delta_1 - \frac{\delta_2}{n^{\times 2}}$$

and solving approximately for $n^\times$

$$n^\times = n + \delta_1 + \frac{\delta_2}{n^2} + \cdots, \tag{5}$$

where δ_1 is an integration constant. $\delta_2 = RbZ^{\times 2}$ is determined by the mechanical system. Of course δ_1 depends also on the second quantum number of the system, for the motion in a central field is double-periodic; one of the periods is the one which has been already considered, namely the motion of the electron from perihelion to perihelion with the main quantum number n; the second is that of the revolution of the perihelion itself with the quantum number k; hk is therefore the angular momentum of the electron and as the rotation of the perihelion is simple-periodic therefore k can change only by ± 1, as in the case of the relativity correction for the H-atom. δ_1 is a function of k; this can also be approximately found by means of relatively simple considerations.

52 PROBLEMS OF ATOMIC DYNAMICS

The expression which we have found in this way for the value of the term

$$\frac{W}{h} = -\frac{RZ^{\times 2}}{(n + \delta_1(k) + \delta_2/n^2 + \cdots)^2} \qquad (6)$$

agrees exactly with the term formulas found empirically by Rydberg (δ_1-term) and Ritz (δ_1- and δ_2-terms).

Since k can have different values, every atom has several series of terms. In fact we should expect, on account of the selection rule for k ($k \to k \pm 1$), that the latter can be so classified that a term of a series can be combined only with the terms of neighboring series. This indeed is the case. It is usual to classify the terms in series according to the following scheme:

$$\begin{array}{ccccc} 1\,s & 2\,s & 3\,s & 4\,s & 5\,s \cdots \\ & 2\,p & 3\,p & 4\,p & 5\,p \cdots \\ & & 3\,d & 4\,d & 5\,d \cdots \\ & & & 4\,f & 5\,f \cdots \end{array}$$

where an s-term can be combined only with a p-term, a p-term with s- and d-terms, etc. From this we conclude with Sommerfeld that the following correspondence holds:

$$\begin{array}{ccccc} & s & p & d & f \\ k = & 1 & 2 & 3 & 4 \end{array}$$

We shall now proceed to determine the main quantum numbers for all the observed terms. For this purpose we must above all determine whether the path in question is a penetrating orbit. We calculate from the observed term $\dfrac{W}{h}$ the effective quantum number $n^\times$ in accordance with the formula

$$n^\times = Z^\times \sqrt{\frac{Rh}{W}}.$$

The aphelion distance then is known, that is the major axis $2\,a^\times$ of the ellipse replacing the path. Moreover the parameter $2\,P$ of this ellipse is known. This parameter depends on the value of k as shown by the formula

$$P = \frac{h^2}{4\pi^2\mu e^2 Z^\times} k^2.$$

Therefore, the whole equivalent ellipse is known approximately and it is possible to determine whether it penetrates into the atomic core, the size of which is known from the ionic volume. If in this way the conclusion is reached that the path is wholly outside the core, then the Rydberg correction δ_1 is small, that is $n^\times$ is nearly an integer. If such is the case then n can be chosen as the next integer to $n^\times$. In fact all the terms corresponding to exterior paths d ($k = 3$), f ($k = 4$) $\cdots$ behave in this manner.

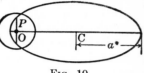

Fig. 10

On the other hand the s-terms ($k = 1$) and the p-terms ($k = 2$) correspond in general to penetrating orbits. Here $n^\times$ departs considerably from integral values. $\delta_1(k)$ is then quite large, frequently larger than 1 or 2. The actual determination of δ_1 requires approximation formulas, for the derivation of which fairly rough assumptions are sufficient. In every case the main quantum number n can be determined with fair certainty.

The main quantum number of the normal state is thus of the greatest interest. The most important result can be stated as follows: For every alkali atom (hydrogen included) the main quantum number of the normal state of the optical electron is increased by 1:

	H	Li	Na	K	Rb	Cs $\cdots$
$n =$	1	2	3	4	5	6 $\cdots$.

LECTURE 8

Bohr's principle of successive building of atoms — Arc and spark spectra — X-ray spectra — Bohr's table of the completed numbers of electrons in the stationary states.

Bohr's construction of the periodic system is based on the supposition that every atom can be derived by the addition of one electron to an ion which is constructed essentially as the previous atom and has the same number of electrons. On this depends the possibility of deriving the structure of one atom from that of the previous one. It is first assumed that the core of the second atom has the same structure as that of the first atom and then, on the basis of a simple estimate of the Rydberg constants, it is seen whether the spectrum is not in contradiction with their value. In many cases we also know the spark spectrum, that is the spectrum of the ionized atom, which is considered as produced by one optical electron rotating around a core of a structure similar to that of the second previous atom having the same number of electrons. We understand from this the so-called "spectroscopic displacement law" given by Sommerfeld and Kossel. The structure of the spectrum of a neutral atom (often called "arc spectrum" because of the most convenient means for its production) resembles the first spark spectrum of the next higher atom, the second spark spectrum of the following atom and so on; except that Rydberg's constant R must be replaced by $4R$, $9R$, $\cdots$ or generally $Z^{\times 2}R$. We have already used the simplest example of this rule, where the correspondence of the spectra is quite exact, when we spoke of the spectra of H, He^+, Li^{++}, $\cdots$ together by introducing an arbitrary nuclear charge Z.

An electronic configuration once formed is buried more and more deeply inside the atom as it proceeds along the periodic system of the elements. Now the X-ray spectra furnish means of examining the inner parts of the atom. The production of these spectra depends, according to Kossel, on the following

THE STRUCTURE OF THE ATOM 55

process: As all the quantum orbits are, so to speak, full, it is impossible for an electron to jump from one orbit to another. It is necessary that an electron be previously removed by supplying energy (electron impact or absorption of X-rays). Then other electrons may fall from higher orbits into the gaps left free and in this way the emission lines of the X-ray spectrum are produced. According to whether the removed electron had the main quantum number $n = 1, 2, 3 \cdots$ we name the line emitted when this electron is replaced, a $K, L, M \cdots$ line; and according to the origin of the substituting electron we indicate the line by indices $K_\alpha, K_\beta \cdots L_\alpha, L_\beta \cdots$ or by new quantum numbers. The correctness of this conception can be tested by observing that for the X-ray lines, the Ritz combination principle must hold. Of course the energy values on the differences of which the frequencies depend are directly given by the so-called *absorption limits*. In the spectrum of absorption of an atom there must exist sharp limits or "edges" which separate the frequencies the energy-quanta $h\nu$ of which are greater or less than the work necessary to remove to infinity the electron describing the orbit which is responsible for the absorption. In this way the system of X-ray terms is determined as exactly as that of the optical spectra.

If we consider the X-ray terms as functions of the atomic numbers Z we obtain in general smooth curves, first discovered by Moseley and Darwin. Only at places where any irregularity in the introduction of electrons occurs are there slight kinks. In this way we can verify the arrangement of the electrons derived from the study of optical spectra. The chief result of this discussion of observed spectra is the following: It is not at all true that all electrons are first introduced in the orbits $n = 1$, then in those for which $n = 2$, $n = 3$, and so on, but on the contrary it is possible that, with electrons already in the orbits $n = 4$, new electrons of higher azimuthal quantum number k fill up an inner shell, for instance $n = 3$. This can be deduced partly from spectroscopic, partly from chemical evidence. If two neighboring elements differ only in that the number of inner electrons, for instance those for which $n = 3$, differ by one, while the number of outer electrons, i.e.

$n = 4$, remains constant (for instance, two), then we should expect that these elements are chemically very similar. We have such groups of similar elements in the fourth period, in the group Sc, Ti, ...Ni, which have the common property of paramagnetism or ferromagnetism. Even more remarkable are the rare earths which are similar in every respect. This is seen in the following presentation of the periodic system of the elements:

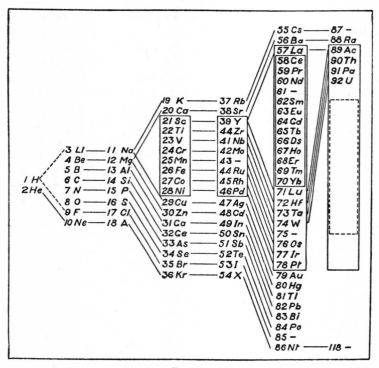

Fig. 11

As a result of all these considerations we give Bohr's table for the arrangement of the electrons:

THE STRUCTURE OF THE ATOM

	1_1	$2_1\ 2_2$	$3_1\ 3_2\ 3_3$	$4_1\ 4_2\ 4_3\ 4_4$	$5_1\ 5_2\ 5_3\ 5_4\ 5_5$	$6_1\ 6_2\ 6_3\ 6_4\ 6_5\ 6_6$	$7_1\ 7_2$
1 H	1						
2 He	2						
3 Li	2	1					
4 Be	2	2					
5 B	2	2 1					
6 C	2	2 (2)					
10 Ne	2	8					
11 Na	2	8	1				
12 Mg	2	8	2				
13 Al	2	8	2 1				
14 Si	2	8	2 (2)				
18 A	2	8	8				
19 K	2	8	8	1			
20 Ca	2	8	8	2			
21 Sc	2	8	8 1	(2)			
22 Ti	2	8	8 2	(2)			
29 Cu	2	8	18	1			
30 Zn	2	8	18	2			
31 Ga	2	8	18	2 1			
36 Kr	2	8	18	8			
37 Rb	2	8	18	8	1		
38 Sr	2	8	18	8	2		
39 Y	2	8	18	8 1	(2)		
40 Zr	2	8	18	8 2	(2)		
47 Ag	2	8	18	18	1		
48 Cd	2	8	18	18	2		
49 In	2	8	18	18	2 1		
54 X	8	8	18	18	8		
55 Cs	2	8	18	18	8	1	
56 Ba	2	8	18	18	8	2	
57 La	2	8	18	18	8 1	(2)	
58 Ce	2	8	18	18 1	8 1	(2)	
59 Pr	2	8	18	18 2	8 1	(2)	
71 Cp	2	8	18	32	8 1	(2)	
72 Hf	2	8	18	32	8 2	(2)	
79 Au	2	8	18	32	18	1	
80 Hg	2	8	18	32	18	2	
81 Tl	2	8	18	32	18	2 1	
86 Nt	2	8	18	32	18	8	
87 —	2	8	18	32	18	8	1
88 Ra	2	8	18	32	18	8	2
89 Ac	2	8	18	32	18	8 1	(2)
90 Th	2	8	18	32	18	8 2	(2)
118 —	2	8	18	32	32	18	8

FIG. 12

We see here how the electronic shell $n = 3$, $k = 1, 2$, which was completed with 8 electrons with Sc ($Z = 21$) begins to increase again for $n = 3$, $k = 3$. The same occurs with Y ($Z = 39$) for $n = 4$, and with La ($Z = 57$) for $n = 5$.

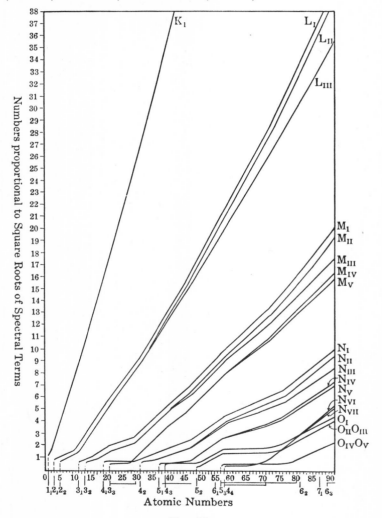

Fig. 13

THE STRUCTURE OF THE ATOM

The X-ray spectra confirm the assumption that internal changes begin with these elements. There are obvious kinks in the curves expressing the relation between X-ray terms and atomic number, which otherwise are quite smooth, for the elements of atomic number $Z = 21, 39, 57$ (Fig. 13).

We may therefore assume that Bohr's arrangement of electronic shells is correct, at any rate as far as the numbers n, k are concerned.

We know nothing about the dynamic mechanism which results in these simple laws. Above all it cannot be explained mechanically why a certain group with a certain main quantum number n is "filled" by a certain limited number of electrons, first 2, then 8, then 18, or why the sub-groups defined by k can also take only a definite number of electrons.

LECTURE 9

Sommerfeld's inner quantum numbers — Attempts toward their interpretation by means of the atomic angular momentum — Breakdown of the classical theory — Formal interpretation of spectral regularities — Stoner's definition of subgroups in the periodic system — Pauli's introduction of four quantum numbers for the electron — Pauli's principle of unequal quantum numbers — Report on the development of the formal theory.

The view of the atom just described has recently led much further in the investigation of the so-called multiplets. Many spectral lines which we have considered here as though they were simple are in fact multiple. For instance, the D-line of sodium is double. Sommerfeld first resolved these lines into terms by introducing a new *inner quantum number j* and giving a selection rule for this number. The possibility of a third quantum number of the optical electron is indicated by the fact that it has *three* degrees of freedom: It need only be supposed that the core is not spherically symmetrical but only symmetrical about an axis. Then the optical electron no longer moves in a central field and therefore the orbit is no longer plane, but, to a first approximation, the motion can be described thus: Assume that the orbit is plane for a single revolution and has the angular momentum k. Then this orbit together with the axis of the atomic core, regarded as a rigid system, is endowed with a precession of angular momentum R around the total momentum J considered fixed in space. K, R, J, as easily shown, are action variables conjugate to the corresponding angles of rotation.

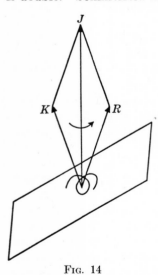

Fig. 14

We place, therefore,

$$K = kh, \quad R = rh, \quad J = jh,$$

where k is the azimuthal quantum number, already introduced above, of the optical electron in its orbit. The quantum number r characterizes the constitution of the core, for, given r and k, j cannot have any value, but only those between $|k - r|$ and $|k + r|$. Also, j, as precessional momentum, can only make jumps

$$j \to j \begin{matrix} \nearrow j - 1 \\ \to j \\ \searrow j + 1 \end{matrix}$$

where the jumps $j \to j \pm 1$ correspond to oscillations of the electric moment perpendicular to the J-axis and $j \to j$ corresponds to oscillations parallel to the J-axis.

We have thus the possibility of explaining multiplets, and the selection rule found empirically by Sommerfeld for the inner quantum number agrees with that found theoretically; but the number of components for given k and r is not verified by experiments. For instance, we are inclined to ascribe to the inert gases, which are certainly highly symmetrical, the angular momentum zero and hence the same to the core of the alkali atoms. But then they should show no separation. If we assume for the inert gases $r = 1$, the values of j lie between $k - 1$ and $k + 1$, therefore $j = k - 1$, $j = k$ and $j = k + 1$. But the alkali atoms have *no* triplets. In the s-states ($k = 1$) they have single lines and in all other states ($k = 2, 3 \cdots$) doublets.

This constitutes a violation of Bohr's principle of selection. The number of possible states of a system consisting of an ion to which an electron has been added is *not* equal to the product of the number of the states of the ion by the number of possible electronic orbits, but one less. Bohr calls this a "non-mechanical constraint" and has repeatedly emphasized that this is a most important deviation from mechanical laws. The difficulty has been overcome formally by introducing half-quantum numbers $\cdots -\frac{3}{2}, -\frac{1}{2}, +\frac{1}{2}, +\frac{3}{2}, \cdots$ and a cabalistic rule.

If a mechanically "logical" system leads to the quantum numbers

$$-3, -2, -1, 0, +1, +2, +3$$

then, by this rule, we replace this series by the one given below:

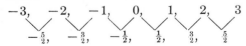

in which the number of terms is one less. In this way the doublets of the alkalies are "explained." The three positions of the core of inert gas type ($r = 1$) with respect to the electronic orbits give only 2 j-values,

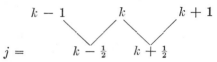

The absolute value of j is of course still arbitrary. Instead of $j = k - \frac{1}{2}, k + \frac{1}{2}$ we can write $j = k - 1, k$, or choose any other normalization which is convenient for the purpose on hand. Only the *number* of possible values of j is important.

The anomalous Zeeman effect is quite similar. Here the classical theory gives for an atom with the total angular momentum $\frac{hj}{2\pi}$, $2j + 1$ orientations in a magnetic field, namely all the values of the magnetic quantum number m between $-j$ and j. In fact, however, only $2j$ terms exist, corresponding to the scheme,

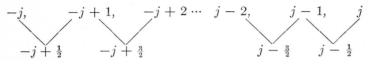

In this way important advances have been made in the systematization of spectra. We shall now review these briefly.

Stoner has given an important generalization of Bohr's theory of the periodic system. He recognized that the electrons in the closed inert gas configurations can be arranged in the following groups:

THE STRUCTURE OF THE ATOM 63

	n	Number of entering electrons
He	1	$2 = 2$
Ne	2	$8 = 2 + 2 + 4$
A	3	$8 = 2 + 2 + 4$
Kr	4	$18 = 2 + 2 + 4 + 4 + 6$
Xe	5	$18 = 2 + 2 + 4 + 4 + 6$
Em	6	$32 = 2 + 2 + 4 + 4 + 6 + 6 + 8$

$$j = 1 \quad 1 \quad 2 \quad 2 \quad 3 \quad 3 \quad 4$$

$$k = 1 \quad 2 \quad 3 \quad 4$$

By dividing the completed number of electrons in each group by 2, the numbers labeled "quantum number j" are obtained. For every $j > 1$, two neighboring values of j are combined to form a larger group with a "quantum number k" equal to the larger j. In this way a one-to-one correspondence is established between quantum numbers and electrons. Stoner was able to show that this development of Bohr's scheme leads to the explanation of many properties of atoms, especially of their spectra in the optical and X-ray regions.

To give only one example: The ionized carbon atom C^+ has a doublet spectrum, which has been analyzed by Fowler. The lines of this spectrum which correspond to jumps to the fundamental orbit give information about this orbit, that is about the normal state of the singly ionized carbon atom C^+. Bohr had originally attributed to the carbon atom four equivalent electrons with $n = 2$, $k = 1$ because chemical facts seemed to demand this equivalence. That would leave to the C^+-ion three equivalent electrons $n = 2$, $k = 1$. Stoner's scheme evidently gives for the carbon atom two electrons $n = 2$, $k = 1$ and two electrons $n = 2$, $k = 2$, of which the former are more strongly attached because they belong to ellipses. The ion C^+ therefore has two electrons $n = 2$, $k = 1$ and one electron $n = 2$, $k = 2$. The jumps of the latter give rise to the spectrum of C^+. The fundamental orbit must hence be shown by its combinations to be a p-term ($k = 2$). In fact observations verify this conclusion and not Bohr's original hypothesis. In the table on p. 57 this result has already been taken account of. We cannot enter into further details here.

Pauli showed Stoner's arrangement to be a consequence of a very general principle. He started from the assumption that spectra behave as though *four* quantum numbers belonged to each electron, that is, that beside the three numbers n, k, j used so far there exists still a fourth m which determines the magnetic separation. Until now j has been interpreted as the resultant of the angular momentum of the electron and the core. Pauli abandons this idea and ascribes all four quantum numbers to *one* electron. The difficulty is that the electron, according to our ordinary ideas, has only three degrees of freedom. It will be seen later that the newest development of the quantum theory seems to lead to a fourth degree of freedom, i. e. an axial rotation, for the electron. Let us forego for the moment a physical explanation and describe briefly Pauli's method. His quantum numbers are normalized somewhat differently from those introduced above. He employs n and k in the usual way, denoting however the latter by k_1, and using instead of j a number k_2 which can *always* have exactly two values $k_2 = k_1 - 1$ and $k_2 = k_1$. Each *single* electron behaves therefore as the optical electron of the alkalies and gives a doublet. The same must hold also for the magnetic quantum number m. For the alkalies m takes, according to observation, $2 k_2$ different values — this can be shown from our cabalistic rule, if k_2 is interpreted as the total momentum. Therefore, all terms belonging to a given k_1 take in all $2 (k_1 - 1) + 2 k_1 = 2 (2 k_1 - 1)$ values. Pauli next observed that Bohr's method of building up atoms by successive steps can be kept in this way. The number of possible states is simply the sum of those of the core and of those of the newly-entering electron (permanence of quantum numbers). If we pass, for instance, from an alkali atom to the neighboring alkaline earth, then the doublet system of the first becomes a system of single and triplet terms. In the singlet system a state with given n, k_1 is decomposed into $1 \times (2 k_1 - 1)$ terms, in the triplet system into $3 (2 k_1 - 1)$ terms. This has been interpreted up to the present as meaning that, in strong fields, there corresponds to the optical electron in spite of mechanics, $2 k_1 - 1$ orientations in every case, while the core is oriented in the single terms along one direction, in the triplet terms along three directions. The latter contradicts the principle of permanence because the free alkali atom can have, in the unexcited state (s-state $k_1 = 1$), only two such orientations. The totality of the $4 (2 k_1 - 1)$ states of the atom can be interpreted as meaning that the core, as in the free alkali atom, can have two states and the optical electron, as in the alkalies, $2 (2 k_1 - 1)$ states. A corresponding explanation can be given in general, but we shall pass over all these details and consider now the connection between Pauli's ideas and Stoner's classification of the periodic system.

Pauli found, that the latter is equivalent to the following general principle: *It never happens that any two electrons in the atom have the same four quantum numbers n, k_1, k_2, m.* If n, k_1, k_2 are given, the number of possible values of m, as was seen above, is $2 k_2$. Therefore, the greatest number of "equivalent" electrons, that is having the same n, k_1, k_2, is also $2 k_2$, otherwise m would be equal for two of these electrons. If the quantum number k_2 is identified with Stoner's j, which also takes for every k (or k_1) the values k, $k - 1$, then Stoner's classification is shown to be a consequence of Pauli's principle. The object of the theory is therefore to understand Pauli's principle, that is either to derive it from the laws of quantum mechanics or to show that it belongs to indemonstrable basic postulates.

Further developments can be briefly described in the following way: According to Pauli electrons aggregate into systems while keeping their own quantum numbers. The energy of such a system, whether a core, a group of outer electrons or a complete atom, depends, however, only on a certain resultant of the quantum numbers of the individual electrons. A distinction must therefore be made between the quantum numbers of the individual electrons and the resultant quantum numbers of the electron groups. If this group is identical with the whole atom, the resulting quantum numbers fix the terms which determine the spectrum. The rules governing the formation of this resultant are mainly of empirical origin. The following is alone mentioned as a theoretical guiding principle. Paschen and Back have discovered that in strong magnetic fields the Zeeman components of a multiplet are displaced with respect to each other so that they correspond to the normal separation (Larmor frequency). This can be interpreted theoretically by assuming that the individual electrons move practically independently of one another in fields where the magnetic energy is much larger than that due to the interaction between the electrons. The electrons therefore precess with normal Larmor frequency. From this follows that in strong fields the magnetic quantum numbers behave additively and the construction of the resultant is referred to this idea.

The development of this conception was very much aided by the investigations of Russell and Saunders. It was already known, according to Götze, that lines appear which correspond to combinations of p-terms with other p'-terms, therefore having the same azimuthal quantum number. Bohr accounted for this by assuming a simultaneous jump of two electrons, whereby the simple harmonic character of the motion, from which we derive the selection rule, $k \to k \pm 1$, is lost. Russell and Saunders found that there exist negative p'-terms, which hence correspond to a state of the atom of higher energy than required for ionization. They were able to explain their observations by introducing a resultant quantum number for the electrons jumping simultaneously and then treating this system with respect to the core in the same way as the optical electron with respect to the core of the alkali atoms. This method was developed systematically by Heisenberg and applied to the practical interpretation of numerous spectra by Hund. The latter succeeded in analyzing completely, among others, the series of the magnetic atoms beginning with scandium and ending with the group iron, cobalt, nickel, deducing not only the completed numbers of the electron groups in the normal states, but also interpreting in a rough way the character of the spectra.

With this we have come to the limit which can be attained by the development of Bohr's fundamental ideas. There is material a plenty. It is now time for the theorist to take the initiative again and lay the foundations of a real dynamics of

THE STRUCTURE OF THE ATOM

atoms. Heisenberg found a short time ago the key to the gate, closed for such a long time, which kept us from the realm of atomic laws. In his brief paper, the leading physical ideas are clearly stated, but only exemplified on account of the lack of appropriate mathematical equipment. The required machinery Jordan and I have discovered in the matrix calculus. Shortly afterwards, as I learned later, Dirac also found an algorithm which is equivalent to ours, but without noticing its identity with the usual mathematical theory of matrices.

LECTURE 10

Introduction to the new quantum theory — Representation of a coördinate by a matrix — The elementary rules of matrix calculus.

In seeking a line of attack for the remodelment of the theory, it must be borne in mind that weak palliatives cannot overcome the staggering difficulties so far encountered, but that the change must reach its very foundations. It is necessary to search for a general principle, a philosophical idea, which has proved successful in other similar cases. We look back to the time before the advent of the theory of relativity, when the electrodynamics of moving bodies was in difficulties similar to those of the atomic theory of today. Then Einstein found a way out of the difficulty by noting that the existing theory operated with a conception which did not correspond to any observable phenomenon in the physical world, the conception of simultaneity. He showed that it is fundamentally impossible to establish the simultaneity of two events occurring in different localities, but rather that a new definition, prescribing a definite method of measurement is required. Einstein gave a method of measurement adapting itself to the structure of the laws of propagation of light and of electromagnetic phenomena in general. Its success justified the method and with it the initial principle involved: *The true laws of nature are relations between magnitudes which must be fundamentally observable.* If magnitudes lacking this property occur in our theories, it is a symptom of something defective. The development of the theory of relativity has shown the fertility of this idea, for the attempt to state the laws of nature in invariant form, independently of the system of coördinates, is nothing but the expression of the desire of avoiding magnitudes which are not observable. A similar situation exists in other branches of physics.

In the case of atomic theory, we have certainly introduced, as fundamental constituents, magnitudes of very doubtful observability, as, for instance, the position, velocity and period of the electron. What we really want to calculate by means of our theory and can be observed experimentally, are the energy levels and the emitted light frequencies derivable from them. The mean radius of the atom (atomic volume) is also an observable quantity which can be determined by the methods of the kinetic theory of gases or other analogous methods. On the other hand, no one has been able to give a method for the determination of the period of an electron in its orbit or even the position of the electron at a given instant. There seems to be no hope that this will ever become possible, for in order to determine lengths or times, measuring rods and clocks are required. The latter, however, consist themselves of atoms and therefore break down in the realm of atomic dimensions. It is necessary to see clearly the following points: All measurements of magnitudes of atomic order depend on indirect conclusions; but the latter carry weight only when their train of thought is consistent with itself and corresponds to a certain region of our experience. But this is precisely not the case for atomic structures such as we have considered so far. I have already called attention to the points where the theory fails.

At this stage it appears justified to give up altogether the description of atoms by means of such quantities as "coördinates of the electrons" at a given time, and instead utilize such magnitudes as are really observable. To the latter belong, besides the energy levels which are directly measureable by electron impacts and the frequencies which are derivable from them and which are also directly measurable, the intensity and the polarization of the emitted waves. We therefore take from now on the point of view that the *elementary waves* are the primary data for the description of atomic processes; all other quantities are to be derived from them. That this standpoint offers more possibilities than the assumption of electronic motions is best understood by considering the Compton effect.

If an X-ray wave of frequency ν impinges on free or loosely bound electrons, it transmits to the latter impacts in every direction. At the same time a secondary X-radiation is emitted having a frequency ν' dependent on the azimuth. According to Compton and Debye this can be quantitatively explained if the energies $h\nu$, $h\nu'$ and the momenta $\dfrac{h\nu}{c}$, $\dfrac{h\nu'}{c}$ are ascribed to the waves and then the laws of conservation of energy and of impulse are applied to the light-quanta and the electrons. But if the process is considered from the point of view of the wave theory, then the change of frequency must be interpreted as a Doppler effect.

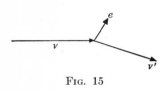

FIG. 15

A calculation of the velocity of the wave-center gives then extremely large values in the direction of the primary X-ray and not in that of the electron. We have therefore struck upon a case in which motion of the electron and motion of the wave-center do not coincide. In the classical theory, where the emitted waves are determined by the harmonic components of the electronic motion, this is of course absolutely unexplainable. We therefore stand before a new fact which forces us to decide whether the electronic motion or the wave shall be looked upon as the primary act. After all theories which postulate the motion have proved unsatisfactory we investigate if this is also the case for the waves.

To begin with consider processes which in the classical theory would correspond to a one-dimensional motion given by a Fourier series for the coördinate q.

$$q(t) = \sum_{\tau} q_{\tau} e^{2\pi i \nu \tau t}. \tag{1}$$

We consider now, not the motion $q(t)$, but the set of all the elementary oscillations

$$q_{\tau} e^{2\pi i \nu \tau t}$$

and try to change them so that they are suitable for the representation, not of the higher harmonics of the motion, but of the real waves of an atom.

Frequencies are *not* therefore in general harmonics ($\nu\tau$), but can be expressed according to Ritz's combination principle as differences of every pair of terms of the series

$$\frac{W_1}{h}, \frac{W_2}{h}, \frac{W_3}{h} \ldots.$$

We therefore write

$$\nu(nm) = \frac{1}{h}(W_n - W_m). \qquad (2)$$

To every jump $n \to m$ corresponds an amplitude and a phase that we denote by the complex amplitude,

$$q(nm) = |q(nm)|e^{i\delta(nm)}. \qquad (3)$$

The set of all possible oscillations is best expressed by ordering them in a square array of terms,

$$\begin{Vmatrix} q(11)e^{2\pi i\nu(11)t} & q(12)e^{2\pi i\nu(12)t} \ldots \ldots \\ q(21)e^{2\pi i\nu(21)t} & q(22)e^{2\pi i\nu(22)t} \ldots \ldots \\ \ldots \ldots \ldots \ldots \ldots \ldots \ldots \ldots \ldots \ldots \ldots \end{Vmatrix}$$

We abbreviate this to

$$q = (q(nm)e^{2\pi i\nu(nm)t}) = (|q(nm)|e^{2\pi i\nu(nm)t + i\delta(nm)}). \qquad (4)$$

In order that this array correspond to a real Fourier series $q(t)$ the condition $\delta(mn) = -\delta(nm)$ must be added, or its equivalent, that $q(nm)$ be transformed into its conjugate $q^*(nm)$ by an interchange of m and n, i.e.,

$$q(mn) = q^*(nm) \qquad (5)$$

because for a real Fourier series the corresponding relation $C_{-\tau} = C^*_\tau$ holds.

The manifold of elementary oscillations is thus naturally represented by a two-dimensional array, while the manifold of harmonics of a motion is represented by a one-dimensional series,

$$C_1 e^{2\pi i\nu t}, \quad C_2 e^{2\pi i 2\nu t}, \quad C_3 e^{2\pi i 3\nu t} \ldots.$$

It is for this reason that, in the theory presented so far, it was necessary to consider simultaneously a whole series of motions, i.e., the stationary states, which are distinguished by another index, i.e., the quantum number n, whereby C and ν become

72   PROBLEMS OF ATOMIC DYNAMICS

functions of n. The array found in this way has neither the correct frequencies nor a simple and unique correspondence to the jumps.

We must now find the laws determining the amplitudes $q(nm)$ and the frequencies $\nu(nm)$. For this purpose we utilize the principle of making the new laws as similar as possible to those of classical mechanics, for the fact that the classical theory of conditioned-periodic motions is in a position to account qualitatively for many quantum phenomena shows that the essential point is not the overthrow of mechanics, but rather a change from classical geometry and kinematics to the new method of representation by means of elementary waves.

As simplest example in classical mechanics we consider the oscillator. We are already familiar with the fact that everything follows once the potential energy $\frac{\kappa}{2} q^2(t)$ is known. The potential energy can also be expressed in terms of elementary waves, because the square of a Fourier series is also a Fourier series

$$q^2(t) = (\Sigma_\tau C_\tau e^{2\pi i \nu \tau t})^2 = \Sigma_\tau D_\tau e^{2\pi i \nu \tau t} \tag{6}$$

where

$$D_\tau = \Sigma_\sigma C_\sigma C_{\tau-\sigma}.$$

The set of quantities D_τ represents therefore the function $q^2(t)$ in quite the same way as the set C_τ represents the function $q(t)$. This can be translated into our square array, as follows: We ask, is it possible to find a multiplication rule for $q(nm)$ by which, out of every array q we can construct a new array which we shall write symbolically q^2, but in which no new frequenci s appear? The latter condition is essential and corresponds to the theorem of classical theory, that the square of a Fourier series, or the product of two such series with the same fundamental frequency, is also a Fourier series with the same fundamental frequency.

This question can be answered by looking upon the square array from the point of view of the mathematician, considering

THE STRUCTURE OF THE ATOM

it as a *matrix* and applying the known rule for the multiplication of matrices. The product of the two matrices

$$a = (a(nm)), \quad b = (b(nm))$$

is defined by the matrix

$$c = (c(nm)) = (\Sigma_k a(nk)b(km)) = ab. \tag{7}$$

If we apply this rule to our array of elementary waves q and multiply it by another array p which has the *same* frequencies $\nu(nm)$ we obtain

$$qp = (\Sigma_k q(nk)e^{2\pi i\nu(nk)t}\, p(km)e^{2\pi i\nu(km)t}),$$

but we have

$$\nu(nk) + \nu(km) = \frac{1}{h}(W_n - W_k) + \frac{1}{h}(W_k - W_m)$$

$$= \frac{1}{h}(W_n - W_m) = \nu(nm).$$

Therefore

$$qp = (\Sigma_k q(nk)p(km)e^{2\pi i\nu(nm)t}); \tag{8}$$

that is, the symbolic product has the same frequencies as its factors. This formula is a profound generalization of the rule for obtaining the Fourier coefficients of the product of two Fourier series. We see that the multiplication rule of matrices is very closely connected with Ritz's principle of combination.

We now give the fundamental rules of matrix calculus. Addition and subtraction are performed by carrying out the required operation on each element:

$$a \pm b = (a(mn) \pm b(mn)). \tag{9}$$

The notation can be simplified further by dropping the factors $e^{2\pi i\nu t}$. The matrix $q = (q(nm))$ represents therefore one coördinate.

The derivative of a matrix with respect to time is the matrix

$$\dot{q} = (2\pi i\nu(nm)q(nm)) \tag{10}$$

where again the exponential factor is dropped. The operation of differentiation can also be expressed in terms of multiplication of matrices. For this purpose we introduce the unit matrix

$$1 = \begin{Vmatrix} 1 & 0 & 0 & \cdots \\ 0 & 1 & 0 & \cdots \\ 0 & 0 & 1 & \cdots \\ \cdots\cdots\cdots\cdots \end{Vmatrix} = (\delta_{nm}), \qquad (11)$$

where

$$\delta_{nm} = \begin{cases} 1 & \text{if } n = m \\ 0 & \text{if } n \neq m. \end{cases}$$

From this we form a diagonal matrix

$$W = (W(nm)) = (W_n \delta_{nm}) = \begin{Vmatrix} W_1 & 0 & 0 & 0 & \cdots \\ 0 & W_2 & 0 & 0 & \cdots \\ 0 & 0 & W_3 & 0 & \cdots \\ \cdots\cdots\cdots\cdots\cdots\cdots \end{Vmatrix} \qquad (12)$$

We now multiply this matrix by the matrix $(q(nm))$. In this connection we note an extremely important theorem in the development of the theory, i.e., that the multiplication of matrices is not commutative. We have

$$Wq = (\sum_k W(nk)q(km)) = (W_n \sum_k \delta_{nk} q(km)) = (W_n q(nm)),$$

but

$$qW = (\sum_k q(nk)W(km)) = (\sum_k q(nk) W_k \delta_{km}) = (W_m q(nm)).$$

If we take the difference,

$$Wq - qW = ((W_n - W_m)q(nm)), \qquad (13)$$

we see that, from Ritz's combination principle,

$$\nu(nm) = \frac{1}{h}(W_n - W_m),$$

follows the formula,

$$\dot{q} = \frac{2\pi i}{h}(Wq - qW). \qquad (14)$$

LECTURE 11

The commutation rule and its justification by a correspondence consideration — Matrix functions and their differentiation with respect to matrix arguments.

We shall now try to translate classical mechanics, as slightly altered as possible, into matrix form. To each coördinate matrix q corresponds a momentum matrix p. We form out of these matrices, by matrix addition and multiplication, in some cases repeated an infinite number of times, the Hamiltonian function H and try to establish the analogue of the canonical differential equations. Here we again encounter the difficulty that products are now non-commutative; qp is not in general equal to pq. At this point the quantum theory makes its appearance. I maintain that the condition

$$pq - qp = \frac{h}{2\pi i} \qquad (1)$$

must be introduced, whereby Planck's constant h is bound up closely with the foundations of the theory. This relation can be made plausible by showing that, in the case of large quantum numbers, it becomes identical with the quantum condition for periodic systems. This limiting case can be described more accurately as follows: We consider large values of m and n, and assume that all $q(mn)$, $p(mn)$ are vanishingly small except if $|m - n| = \tau$ is small compared with m and n. For simplicity we consider only the case where $p = \mu \dot{q}$, therefore

$$p(mn) = 2\pi i \mu \nu(mn) q(mn).$$

Let us consider especially the diagonal elements of our quantum condition (1)

$$\sum_k (p(nk)q(kn) - q(nk)p(kn)) = \frac{h}{2\pi i} \qquad (2)$$

or

$$\sum_k \nu(nk)|q(nk)|^2 = -\frac{h}{8\pi^2 \mu}$$

PROBLEMS OF ATOMIC DYNAMICS

for which we can write

$$\sum_{\tau>0} (\nu(n,n+\tau)|q(n,n+\tau)|^2 + \nu(n,n-\tau)|q(n,n-\tau)|^2) = -\frac{h}{8\pi^2\mu}$$

or, since $\nu(mn) = -\nu(nm)$,

$$\sum_{\tau>0} (\nu(n+\tau,n)|q(n+\tau,n)|^2 - \nu(n,n-\tau)|q(n,n-\tau)|^2) = \frac{h}{8\pi^2\mu}.$$

If we place

$$f_\tau(n) = \nu(n, n-\tau)|q(n, n-\tau)|^2.$$

we may write

$$\sum_{\tau>0} \tau \cdot \frac{f_\tau(n+\tau) - f_\tau(n)}{\tau} = \frac{h}{8\pi^2\mu}.$$

If we pass to the limit $n >> \tau$, we obtain the classical formulas. Placing $nh = I$, we obtain

$$\nu(n, n-\tau) = \tau \frac{W(n) - W(n-\tau)}{\tau h} \rightarrow \tau \frac{dW}{dI} = \tau\nu \quad (3)$$

which is the classical frequency of the τth harmonic. Further, the corresponding amplitude is

$$q(n, n-\tau) \rightarrow q_\tau(I).$$

Therefore

$$f_\tau(n) \rightarrow f_\tau(I) = \nu\tau|q_\tau(I)|^2$$

and

$$\frac{1}{8\pi^2\mu} = \sum_{\tau>0} \tau \frac{f_\tau(n+\tau) - f_\tau(n)}{h\tau} \rightarrow \sum_{\tau>0} \tau \frac{\partial}{\partial I} f_\tau(I). \quad (4)$$

This formula, however, is the quantum condition of Bohr's theory

$$\int_0 pdq = I = hn,$$

for if we set

$$q(t) = \sum_\tau q_\tau e^{2\pi i\nu\tau t}$$

we obtain

$$I = \mu \int_0^{\frac{1}{\nu}} \dot{q}^2 dt = -\mu(2\pi)^2 \int_0^{\frac{1}{\nu}} \sum_{\tau,\sigma=-\infty}^{\infty} \nu\sigma q_\tau q_\sigma e^{2\pi i \nu(\tau+\sigma)t} dt$$
$$= 4\pi^2\mu \cdot 2 \sum_{\tau>0} \tau^2 \nu q_\tau q_{-\tau} = 8\pi^2\mu \sum_{\tau>0} \tau \cdot \nu\tau |q_\tau|^2$$

and differentiating with respect to I

$$\frac{1}{8\pi^2\mu} = \sum_{\tau>0} \tau \frac{\partial}{\partial I} \nu\tau |q_\tau|^2 = \sum_{\tau>0} \tau \frac{\partial}{\partial I} f_\tau(I) \tag{5}$$

in agreement with the limit given above.

These correspondence considerations justify in a certain sense the diagonal elements of the fundamental relation (1). In order to approach as closely as possible to commutativity it is reasonable to set all elements except those on the diagonal equal to zero. Owing to this commutation law, calculations with matrices become determinate. We can therefore construct functions of p and q by repeated multiplications and additions.

We have for instance the energy-function of the harmonic oscillator (Mass = μ):

$$H = \frac{1}{2\mu} p^2 + \frac{\kappa}{2} q^2. \tag{6}$$

To form the canonical equations we must first introduce the operation of differentiation. The derivative of a matrix function $f(x)$ with respect to the argument-matrix x is defined by

$$\frac{df}{dx} = \lim_{\alpha \to 0} \frac{f(x+\alpha) - f(x)}{\alpha} \tag{7}$$

where $\alpha(mn)$ is the product of the unit matrix by a number α

$$\alpha(mn) = \alpha \delta_{mn}.$$

The multiplication by such a matrix, or its reciprocal

$$\alpha^{-1}(mn) = \frac{1}{\alpha} \delta_{mn}$$

PROBLEMS OF ATOMIC DYNAMICS

is commutative and therefore our definition has a unique meaning. We have, for instance,

$$\frac{dx}{dx}(mn) = \lim_{\alpha \to 0} \frac{1}{\alpha}[x(mn) + \alpha\delta_{mn} - x(mn)] = \delta_{mn},$$

that is

$$\frac{dx}{dx} = 1.$$

Similarly

$$\frac{dx^2}{dx}(mn) = \lim_{\alpha \to 0} \frac{1}{\alpha}[\Sigma_k (x_{mk} + \alpha\delta_{mk})(x_{kn} + \alpha\delta_{kn}) - \Sigma_k x_{mk}x_{kn}] = 2x_{mn},$$

that is

$$\frac{dx^2}{dx} = 2x.$$

The product rule

$$\frac{d}{dx}(\phi\psi) = \phi\frac{d\psi}{dx} + \frac{d\phi}{dx}\psi \tag{8}$$

is proved as in ordinary calculus:

$$\frac{d}{dx}(\phi\psi) = \lim_{\alpha \to 0}\frac{1}{\alpha}\Big[\phi(x+\alpha)\psi(x+\alpha) - \phi(x)\psi(x)\Big]$$

$$= \lim_{\alpha \to 0}\frac{1}{\alpha}\Big[\phi(x+\alpha)\psi(x+\alpha) - \phi(x+\alpha)\psi(x)$$

$$+ \phi(x+\alpha)\psi(x) - \phi(x)\psi(x)\Big]$$

$$= \phi\frac{d\psi}{dx} + \frac{d\phi}{dx}\psi$$

where it should be observed that the order ϕ, ψ must be conserved. From this we deduce at once

$$\frac{dx^n}{dx} = nx^{n-1},$$

whence it follows that all the rules of ordinary differential calculus hold. The partial derivative of a matrix function of several argument-matrices $f(x_1, x_2 \cdots)$ with respect to one of them, say x_1, is obtained by applying our definition of differentiation to x_1 only, while x_2, $x_3 \cdots$ are held constant.

LECTURE 12

The canonical equations of mechanics — Proof of the conservation of energy and of the "frequency condition" — Canonical transformations — The analogue of the Hamilton-Jacobi differential equation.

We can now write the *canonical equations*

$$\left.\begin{array}{r}\dot{q} = \dfrac{\partial H}{\partial p} \\[6pt] \dot{p} = -\dfrac{\partial H}{\partial q}\end{array}\right\} \quad (1)$$

They form in reality an infinite number of equations for an infinite number of unknowns, for the matrices on the right and left-hand sides must be equal element by element.

To establish the law of conservation of energy we need the following lemmas: Let $f(qp)$ be any matrix function of p and q. Then

$$\left.\begin{array}{r}fq - qf = \dfrac{h}{2\pi i}\dfrac{\partial f}{\partial p} \\[6pt] pf - fp = \dfrac{h}{2\pi i}\dfrac{\partial f}{\partial q}\end{array}\right\} \quad (2)$$

To prove these relations we first assume that they are true for any two given functions ϕ and ψ and show that they are also true for $\phi + \psi$ and $\phi\psi$. For $\phi + \psi$ this is trivial, for $\phi\psi$ a simple calculation gives

$$\phi\psi q - q\phi\psi = \phi(\psi q - q\psi) + (\phi q - q\phi)\psi$$
$$= \frac{h}{2\pi i}\left(\phi\frac{\partial \psi}{\partial p} + \frac{\partial \phi}{\partial p}\psi\right) = \frac{h}{2\pi i}\frac{\partial}{\partial p}\phi\psi$$

and an analogous relation for $p\phi\psi - \phi\psi p$. But our relations hold for $f = p$ and $f = q$ and therefore hold for every function, as functions have already been defined by repeated application of the elementary operations.

By Equations (14), Lecture 10, and (2), this Lecture, we may write the canonical equations (1)

$$Wq - qW = Hq - qH \brace Wp - pW = Hp - pH \quad (3)$$

or

$$(W - H)q - q(W - H) = 0$$
$$(W - H)p - p(W - H) = 0.$$

$W - H$ is therefore commutable with p and q, hence also with any function of p and q, in particular with $H(pq)$. We thus have

$$(W - H)H - H(W - H) = 0$$

or

$$WH - HW = 0.$$

From this follows, by Equation (14), Lecture 10,

$$\dot{H} = 0 \quad (4)$$

which proves the *conservation of energy*. H is thus seen to be a diagonal matrix

$$H(nm) = \begin{cases} H_n & \text{for } n = m \\ 0 & \text{for } n \neq m \end{cases} \quad (5)$$

For the elements, the first of Equations (3) can be written,

$$q(nm)(W_n - W_m) = q(nm)(H_n - H_m).$$

Therefore,

$$H_n - H_m = W_n - W_m = h\nu(nm) \quad (6)$$

whence *Bohr's frequency condition* follows as a consequence of our postulates. By a suitable choice of an arbitrary constant we can place

$$H_n = W_m \quad (7)$$

and this gives to the Ritz combination principle the more precise meaning of the Einstein-Bohr frequency condition.

The whole proof can also be reversed. **We know that the**

principle of conservation of energy and the frequency condition are correct. If, therefore, the energy-function H is given as an analytic function of any two variables P, Q, then, provided that

$$PQ - QP = \frac{h}{2\pi i},$$

the canonical equations

$$\dot{Q} = \frac{\partial H}{\partial P}, \qquad \dot{P} = -\frac{\partial H}{\partial Q}$$

hold. This is true because the expressions $HP - PH$ and $HQ - QH$ can always, as we have shown, be interpreted in two ways, either as partial derivatives of H or, as H is constant, as derivatives of Q or P with respect to time. Therefore, we understand by a *canonic transformation* $pq \rightarrow PQ$ one for which

$$pq - qp = PQ - QP = \frac{h}{2\pi i} \qquad (8)$$

for then the canonical equations hold for p, q as well as for P, Q.

A general transformation which satisfies this condition is

$$\left.\begin{array}{l} P = SpS^{-1} \\ Q = SqS^{-1} \end{array}\right\} \qquad (9)$$

where S is any arbitrary matrix. Probably, this is the most general canonic transformation. It has the simple property that for any function $f(PQ)$ the relation

$$f(PQ) = Sf(pq)S^{-1} \qquad (10)$$

holds, where $f(pq)$ is formed from $f(PQ)$ by replacing P by p and Q by q without changing the form of the function. We shall show that if this theorem is true for two functions ϕ, ψ, it is also true for $\phi + \psi$ and $\phi\psi$. For $\phi + \psi$ it is evident. For $\phi\psi$ we have

$$\phi(PQ)\psi(PQ) = S\phi(pq)S^{-1}S\psi(pq)S^{-1} = S\phi(pq)\psi(pq)S^{-1}.$$

As the proposition holds for $f = p$ or $f = q$, it holds in general for all analytic functions.

The importance of the canonic transformations is based on

the following theorem: If any pair of variables p_0, q_0 is given which satisfies the condition

$$p_0 q_0 - q_0 p_0 = \frac{h}{2\pi i}$$

we can reduce the problem of integrating the canonical equations for an energy function $H(pq)$ to the following one: A function S is to be determined, such that

$$H(pq) = SH(p_0 q_0)S^{-1} = W \tag{11}$$

becomes a diagonal matrix. Then the solution of the canonical equations has the form

$$p = S p_0 S^{-1}, \qquad q = S q_0 S^{-1}.$$

We have therefore a complete analogue of *Hamilton-Jacobi's differential equation*. S corresponds to the action-function.

LECTURE 13

The example of the harmonic oscillator — Perturbation theory.

Let us now illustrate these abstract considerations by an example. For this purpose we choose the harmonic oscillator, for which

$$H = \frac{p^2}{2\mu} + \frac{\kappa}{2}q^2. \tag{1}$$

The canonical equations

$$\dot{q} = \frac{p}{\mu}, \qquad \dot{p} = -\kappa q \tag{2}$$

give by elimination of p and placing $\frac{\kappa}{\mu} = (2\pi\nu_0)^2$

$$\ddot{q} + (2\pi\nu_0)^2 q = 0 \tag{3}$$

or, more explicitly,

$$[\nu^2(nm) - \nu_0^2]q(nm) = 0. \tag{4}$$

To this is added the commutation relation which gives

$$\sum_k [\nu(nk) - \nu(km)]q(nk)q(km) = \left\{ \begin{array}{l} -\dfrac{h}{4\pi^2\mu} \text{ if } n = m \\ 0 \quad\quad \text{ if } n \neq m \end{array} \right\} \tag{5}$$

There follows from the equation of motion that $q(nm)$ can differ from zero only if

$$\nu(nm) = \frac{1}{h}(W_n - W_m) = \pm\nu_0. \tag{6}$$

In the row m of the matrix there are therefore at most two non-vanishing elements, i.e., those for which

$$W_n = W_m + h\nu_0 \quad \text{or} \quad W_n = W_m - h\nu_0.$$

Evidently the order of the elements in the diagonal of a matrix is of no importance. If we perform the same permutation on the rows and columns, all matrix equations are unaltered. We can therefore choose $W_m = W_0$ arbitrarily and denote the "neighboring values" $W_0 + h\nu_0$ and $W_0 - h\nu_0$ of W_0 by the symbols W_1 and W_{-1}. Each of these has again neighboring values which differ from it by $h\nu_0$, etc. In this way we obtain an arithmetical series of energy levels,

$$W_n = W_0 \pm nh\nu_0. \tag{7}$$

The diagonal elements of the commutation relation (5) give

$$\frac{h}{8\pi^2\mu} = -\sum_k \nu(nk)|q(nk)|^2 = \nu_0[|q(n, n+1)|^2 - |q(n, n-1)|^2]. \tag{8}$$

Whence it follows that $|q(n, n+1)|^2$ also form an arithmetical series with the difference $h/8\pi^2\mu\nu_0$. Since all these terms are positive, the series must stop somewhere. We have therefore

$$|q(1, 0)|^2 = \frac{h}{8\pi^2\mu\nu_0}$$

$$|q(n+1, n)|^2 = |q(n, n-1)|^2 + \frac{h}{8\pi^2\mu\nu_0}, \quad n = 1, 2, 3\ldots$$

therefore

$$|q(n+1, n)|^2 = (n+1)\frac{h}{8\pi^2\mu\nu_0}. \tag{9}$$

It is apparent at once that all other elements of the matrix $pq - qp$ are actually zero. We verify further the conservation of energy:

$$H(nm) = 4\pi^2\mu\sum_k[\nu_0^2 - \nu(nk)\nu(km)]q(nk)q(km). \tag{10}$$

This vanishes for $n \neq m$ and we have

$$H(nn) = 4\pi^2\mu\nu_0^2[|q(n+1, n)|^2 + |q(n, n-1)|^2]$$
$$= h\nu_0 \tfrac{1}{2}(2n+1) = h\nu_0(n+\tfrac{1}{2}). \tag{11}$$

The quantity W_0 introduced above has therefore the value $\tfrac{1}{2}h\nu_0$. The energy at absolute zero, which has been considered already by Planck and Nernst in statistical problems of the quantum theory, appears here quite naturally.

THE STRUCTURE OF THE ATOM 85

The formula for the complex amplitudes

$$q(n+1, n)e^{2\pi i\nu_0 t} = \sqrt{\frac{h}{8\pi^2\mu\nu_0}(n+1)}e^{i(2\pi\nu_0 t+\phi_n)} \qquad (12)$$

involves arbitrary phases ϕ_n which are of great importance for the statistical behavior of the resonator. Besides, Equation (12) goes over into the classical formula

$$q(t) = \sqrt{\frac{I}{8\pi^2\mu\nu_0}}e^{i(2\pi\nu_0 t+\phi)}, \quad I = hn \qquad (13)$$

for large values of n.

The theory of the harmonic oscillator can be used as a starting point for the calculation of more general systems, if we consider these as derived from the former by variation of one of its parameters. The required process can be developed in a way closely analogous to the classical perturbation theory.

We assume the energy given as a power series in the parameter λ,

$$H = H_0(pq) + \lambda H_1(pq) + \lambda^2 H_2(pq) + \cdots. \qquad (14)$$

Let the mechanical problem defined by $H_0(pq)$ be solved. We know the solution p_0, q_0 which satisfies the condition

$$p_0 q_0 - q_0 p_0 = \frac{h}{2\pi i}.$$

and for which $H_0(p_0 q_0)$ becomes a diagonal matrix W^0. Now we try to determine a transformation S such that if

$$p = Sp_0 S^{-1}, \quad q = Sq_0 S^{-1}, \qquad (15)$$

$H(pq)$ is transformed into a diagonal matrix W. This means that S satisfies the Hamilton-Jacobi equation

$$H(pq) = SH(p_0 q_0)S^{-1} = W. \qquad (16)$$

To solve this equation we place

$$\left.\begin{array}{l} W = W^0 + \lambda W^{(1)} + \lambda^2 W^{(2)} + \cdots \\ S = 1 + \lambda S_1 + \lambda^2 S_2 + \cdots. \end{array}\right\} \qquad (17)$$

Then we have

$$S^{-1} = 1 - \lambda S_1 + \lambda^2(S_1^2 - S_2) - \cdots + \cdots.$$

Substituting in Equation (16),

$$(1 + \lambda S_1 + \lambda^2 S_2 + \cdots)(H_0(p_0q_0) + \lambda H_1(p_0q_0) + \lambda^2 H_2(p_0q_0) + \cdots)$$
$$(1 - \lambda S_1 + \lambda^2(S_1^2 - S_2) + \cdots) = W^0 + \lambda W^{(1)} + \lambda^2 W^{(2)} + \cdots$$

and equating the coefficients of like powers of λ we obtain the following system of approximate equations:

$$\left.\begin{aligned} H_0(p_0q_0) &= W^0 \\ S_1H_0 - H_0S_1 + H_1 &= W^{(1)} \\ S_2H_0 - H_0S_2 + H_0S_1^2 - S_1H_0S_1 + S_1H_1 - H_1S_1 + H_2 &= W^{(2)} \\ \cdots\cdots\cdots\cdots\cdots\cdots\cdots\cdots\cdots\cdots\cdots\cdots\cdots\cdots\cdots\cdots\cdots& \\ S_rH_0 - H_0S_r + F_r(H_0, \cdots H_r, S_0, \cdots S_{r-1}) &= W^{(r)} \end{aligned}\right\} \quad (18)$$

where $H_0, H_1 \cdots$ are to be considered as functions of p_0, q_0.

The first equation is satisfied. The others can be solved successively in a way quite analogous to that used in classical theory: The average of the energy is first formed in order to fix the energy constant, for

$$S_rH_0 - H_0S_r = -(W^0S_r - S_rW^0)$$

has no diagonal terms. There follows, in general,

$$W^{(r)} = \overline{F_r}, \quad \text{i.e.} \quad W_n^{(r)} = F_r(nn).$$

We have, further,

$$W_n^0 S_r(mn) - W_m^0 S_r(mn) + F_r(mn) = 0, \ m \neq n$$

or

$$S_r(mn) = \frac{F_r(mn)}{h\nu_0(mn)}(1 - \delta_{mn}) \tag{19}$$

where $\nu_0(mn)$ are the frequencies of the unperturbed motion.

This solution satisfies the condition

$$S\widetilde{S}^* = 1 \tag{20}$$

where the symbol $\smile$ denotes transposition of the rows and columns and the symbol* the substitution of conjugate complex quantities. As S is only obtained by successive calculation of the approximations $S_1, S_2 \cdots$ this relation can be proved only by successive steps. We shall restrict ourselves to the first

step. If we must have
$$S\widetilde{S}^* = (1 + \lambda S_1 + \cdots)(1 + \lambda \widetilde{S}_1^* + \cdots) = 1$$
then
$$S_1 + \widetilde{S}_1^* = 0,$$
but our general formula (19) gives
$$S_1(mn) = \frac{H_1(mn)}{h\nu_0(mn)}(1 - \delta_{mn}),$$
therefore
$$\widetilde{S}_1^*(mn) = S_1^*(nm) = \frac{H_1^*(nm)}{h\nu_0(nm)}(1 - \delta_{mn}). \qquad (21)$$

As H_1 is an Hermitian matrix, that is since
$$H_1^*(nm) = H_1(mn),$$
it follows that
$$\widetilde{S}_1^*(mn) = \frac{H_1(mn)}{-h\nu_0(mn)}(1 - \delta_{mn}) = -S_1(mn).$$

The importance of the relation $S\widetilde{S}^* = 1$ arises from the fact that the Hermitianness of the matrices p, q is a consequence of this relation. The rule
$$\widetilde{(ab)} = \widetilde{b}\,\widetilde{a} \qquad (22)$$
holds, as can be easily deduced from the definition of the products:
$$\sum_k a(nk)b(km) = \sum_k \widetilde{b}(mk)\widetilde{a}(kn).$$
From this follows that
$$q^* = S^* q_0^* (S^*)^{-1} = \widetilde{S}^{-1}\widetilde{q}_0\widetilde{S} = \widetilde{q} \qquad (23)$$
and similarly for p.

If we place,
$$\left.\begin{array}{l}q = q_0 + \lambda q_1 + \cdots = (1 + \lambda S_1 + \cdots) q_0 (1 - \lambda S_1 + \cdots) \\ p = p_0 + \lambda p_1 + \cdots = (1 + \lambda S_1 + \cdots) p_0 (1 - \lambda S_1 + \cdots)\end{array}\right\}$$

then we have, as a first approximation,
$$q_1 = S_1 q_0 - q_0 S_1,$$
$$p_1 = S_1 p_0 - p_0 S_1.$$

Or more explicitly,

$$\left. \begin{aligned} q_1(mn) &= \frac{1}{h} \sum_k{}' \left(\frac{H_1(mk)q_0(kn)}{\nu_0(mk)} - \frac{q_0(mk)H_0(kn)}{\nu_0(kn)} \right) \\ p_1(mn) &= \frac{1}{h} \sum_k{}' \left(\frac{H_1(mk)p_0(kn)}{\nu_0(mk)} - \frac{p_0(mk)H_0(kn)}{\nu_0(kn)} \right). \end{aligned} \right\} \quad (24)$$

For the energy we obtain, as a second approximation,
$$W^{(2)} = \overline{H_0 S_1{}^2} - \overline{S_1 H_0 S_1} + \overline{S_1 H_1} - \overline{H_1 S_1} + \overline{H_2},$$
or

$$\left. \begin{aligned} W_n^{(2)} &= \Sigma'_k (W_n^0 S_1(nk) S_1(kn) - S_1(nk) S_1(kn) W_k^0 \\ &\quad + S_1(nk) H_1(kn) - H_1(nk) S_1(kn)) + H_2(nn) \\ W_n^{(2)} &= H_2(nn) + \frac{1}{h} \sum_k{}' \frac{H_1(nk)H_1(kn)}{\nu_0(nk)}. \end{aligned} \right\} \quad (25)$$

LECTURE 14

The meaning of external forces in the quantum theory and corresponding perturbation formulas — Their application to the theory of dispersion.

Before discussing the significance of these formulas, we consider the more general case where the Hamiltonian function contains the time t explicitly. This can be easily taken account of formally by introducing in $H(t, p, q)$ instead of t a new coördinate q^0, to which corresponds a momentum p^0, and considering the Hamiltonian function

$$H^\times = H(q^0, p, q) + p^0. \tag{1}$$

To q^0, p^0 correspond the canonical equations,

$$\dot{q}^0 = \frac{\partial H^\times}{\partial p^0} = 1, \qquad \dot{p}^0 = -\frac{\partial H^\times}{\partial q^0} = -\frac{\partial H}{\partial t} \tag{2}$$

of which the first says that q^0 is the time and the second defines p^0.

A closer consideration leads to an important difficulty. The introduction of a function H depending explicitly on t has evidently the physical meaning that the reaction of the system A in question on other systems B which act on A is so small that it can be neglected, and that the quantities depending on these external systems B can be considered to be the same functions of time as they would be without the presence of A. In classical theory, where the interactions of two systems depend only on their instantaneous motion, the condition for this is that the coupling energy be small. But in the quantum theory this is not obviously so. Here the reaction depends, as our perturbation formulas show, not only on the instantaneous state of the system, but on all the states of the system together, for the products occurring in the formulas contain sums over all the states. The perturbation of the system A, due to a motion of the system B given as a function of the time can be taken care of only as long as approximations are restricted to those for which the quantities belonging to B enter only linearly in the perturbation function H_1. Higher approximations have no meaning even in the case of a weak coupling. But if the assumption is made that the system A under consideration is negligible energetically compared with the external systems B, then going over to higher approximations can also be justified in the quantum theory.

We shall restrict ourselves here to the first approximation q_1, p_1. We consider the special case where the system defined by H_0 is acted upon by an electric field **E**. Then the perturbation function is, to a first approximation,

$$H_1 = eq_0E. \quad (3)$$

According to what has been said above, **E** can be looked upon as a function of the time. If in particular we are considering a monochromatic light wave of the frequency ν

$$E = E_0 \cos 2\pi\nu t,$$

therefore

$$H_1 = eE_0q_0 \cos 2\pi\nu t = \tfrac{1}{2} eE_0q_0(e^{2\pi i\nu t} + e^{-2\pi i\nu t})$$

then we get for the perturbation of the coördinates,

$$q_1(mn) = \frac{E_0 e}{2h} \sum_k \left(\frac{q_0(mk)q_0(kn)}{\nu_0(mk) + \nu} - \frac{q_0(mk)q_0(kn)}{\nu_0(kn) + \nu} \right)$$

or, as $p_1 = \mu\dot{q}_1$,

$$q_1(mn) = \frac{E_0 e}{2h \cdot 2\pi i\mu} \sum_k \frac{q_0(mk)p_0(kn) - p_0(mk)q_0(kn)}{(\nu_0(mk) + \nu)(\nu_0(kn) + \nu)}. \quad (4)$$

For the diagonal terms we have, in particular,

$$q_1(nn) = -\frac{E_0 e}{2h \cdot 2\pi i\mu} \sum_k \frac{q_0(nk)p_0(kn) - p_0(nk)q_0(kn)}{\nu_0^2(nk) - \nu^2}. \quad (5)$$

The polarization produced by the field **E** is obtained by multiplying q_1 by the charge e and then the index of refraction can be calculated by well-known methods.

This formula for $q_1(nn)$ contains Kramers' theory of dispersion, which was found by considerations of correspondence. To understand its meaning we recall the relation between the

THE STRUCTURE OF THE ATOM 91

theory of dispersion and the quantum theory of multiple-periodic systems. When a light wave acts on such a system, the electronic orbits perform oscillations. The resonance points of these forced oscillations lie evidently where the Fourier analysis of the orbits leads to a harmonic overtone. Debye attempted to calculate the dispersion formula for the hydrogen molecule using the model shown in Fig. 16, and Sommerfeld extended this process to more general molecular models with electrons arranged in rings. If they found a fairly good agreement with measurements of refractive indices it was only because the range of measurements lay very far from the characteristic resonance points. The incorrectness of the formula follows already from the fact that some of the resonance points have imaginary proper frequencies, which is always a sign of instability of the motion. It is rendered more evident by the fact that the resonance points have no relation to the frequencies which the system would emit according to

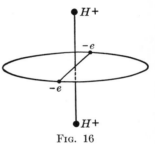

Fig. 16

the quantum theory. It is quite clear, however, that the frequencies actually emitted must determine essentially the resonance or dispersion curve, and not the higher harmonics of the stationary motion which are not optically observable.

The first step towards a rational change of the theory of dispersion in this sense was made by Ladenburg. His dispersion formula consists essentially of those terms in the above expression (5) for $q_1(nn)$ for which $n < k$ and which, therefore, correspond to "upward jumps," that is, to absorption processes. Ladenburg also discovered a relation between the numerator of the dispersion formula $|q_0(nk)|^2 \nu_0(kn)$ and the transition probabilities between the states n and k which appear in Einstein's derivation of Planck's formula.

Kramers has given the complete expression for $q_1(nn)$ in which the emission terms $(n > k)$ are also taken into account. Since

$$\nu_0(kn) = -\nu_0(nk)$$

these terms give "negative" contributions to the dispersion. Kramers' formula has the great advantage of reducing in the limit to the classical formula for the influence of an alternating field on a multiple-periodic system. It therefore satisfies the principle of correspondence.

The case of a constant electric field (Stark effect), as represented by our original formula, was used by Pauli to estimate the intensity of the spectral lines of the mercury atom which do not appear in its natural state (for which $q_0(nm) = 0$) and which are first excited by the field ($q_1(nm) \neq 0$).

Kramers' work suggested to me that quite generally the perturbation energy cannot depend on the classical frequencies of the unperturbed system, but rather on the quantic frequencies and this has been recently confirmed by Schrödinger's considerations on the actual structure of certain line spectra, e.g. aluminum. By correspondence considerations I arrived at the expression (25) Lecture 13 for the perturbation energy $W^{(2)}$. By similar considerations Heisenberg and Kramers also found and discussed the expression $q_1(nm)$ [Eq. (4), Lecture 14] for a light wave. They correspond to the phenomenon that light of frequency ν is not only scattered as light of the same frequency, as in the classical theory, but also as light of other colors belonging to the combination frequencies $\nu \pm \nu_0(nk)$. This phenomenon had been already postulated by Smekal from considerations on light quanta.

Consider finally the limiting case of very high frequencies of the exciting light

$$\nu >> |\nu_0(mk)|, \quad \nu >> |\nu(kn)|.$$

We then obtain

$$q_1 = - \frac{E_0 e}{2 h \cdot 2 \pi i \nu^2 \mu} (p_0 q_0 - q_0 p_0),$$

and since

$$p_0 q_0 - q_0 p_0 = \frac{h}{2 \pi i}$$

therefore

$$q_1 = \frac{E_0 e}{8 \pi^2 \nu^2 \mu}. \tag{6}$$

THE STRUCTURE OF THE ATOM 93

Compare this with the excitation of a free electron by the same electric field $E_0 \cos 2\pi\nu t$. Here we must only take that part $\frac{1}{2} E_0 e^{2\pi i\nu t}$ corresponding to an element of the matrix. We have the differential equation

$$\mu \ddot{q}_1 = \tfrac{1}{2} e E_0 e^{2\pi i\nu t}$$

the solution of which is

$$q_1 = \frac{E_0 e}{8 \pi^2 \nu^2 \mu}.$$

Our quantic commutation rule can therefore be interpreted as the condition that, for sufficiently high frequencies, the electron behave as in classical theory, where the scattered light of frequency ν has the correct intensity and the scattered light of the combination frequencies vanishes. Starting from this condition Kuhn and Thomas found a formula equivalent to the commutation rule, as already stated above, and they and Reiche applied it to dispersion problems.

LECTURE 15

Systems of more than one degree of freedom — The commutation rules — The analogue of the Hamilton-Jacobi theory — Degenerate systems.

We now consider systems of f degrees of freedom. By an immediate generalization they may be represented by $2f$-dimensional matrices

$$\left.\begin{array}{l} q_k = (q_k(n_1, n_2 \cdots n_f;\ m_1, m_2 \cdots m_f)) \\ p_k = (p_k(n_1, n_2 \cdots n_f;\ m_1, m_2 \cdots m_f)) \end{array}\right\} \quad (1)$$

This representation is sometimes very convenient and clear, but not at all necessary. We can always imagine the matrix written in two dimensions. Then, as already shown for one degree of freedom, the expression of the stationary states as given by the arrangement of the rows is quite immaterial, as contrasted with the older theory. We can therefore always transform a $2f$-dimensional matrix into a two-dimensional one. We can for instance write a 4-dimensional matrix $q(n_1, n_2;\ m_1, m_2)$ as follows:

$$q = \left\| \begin{array}{llll} q(1,1;1,1) & q(1,1;1,2) \cdots & q(1,1;2,1) & q(1,1;2,2) \cdots \\ q(1,2;1,1) & q(1,2;1,2) \cdots & q(1,2;2,1) & q(1,2;2,2) \cdots \\ \cdots\cdots\cdots\cdots\cdots\cdots\cdots\cdots\cdots\cdots\cdots\cdots\cdots\cdots\cdots \\ q(2,1;1,1) & q(2,1;1,2) \cdots & q(2,1;2,1) & q(2,1;2,2) \cdots \\ q(2,2;1,1) & q(2,2;1,2) \cdots & q(2,2;2,1) & q(2,2;2,2) \cdots \\ \cdots\cdots\cdots\cdots\cdots\cdots\cdots\cdots\cdots\cdots\cdots\cdots\cdots\cdots\cdots \end{array} \right\|$$

The definitions of addition and of multiplication are quite independent of the order of the indices. The rules of matrix calculus are therefore applicable as before. We can therefore define a Hamiltonian function

$$H(q_1 \cdots q_f,\ p_1 \cdots p_f)$$

and have the equations of motion:

$$\dot{q}_k = \frac{\partial H}{\partial p_k}, \qquad \dot{p}_k = -\frac{\partial H}{\partial q_k}. \quad (2)$$

THE STRUCTURE OF THE ATOM   95

The *quantic commutation rules* are fundamental. We make the following immediate generalization:

$$\left.\begin{array}{l} p_k q_l - q_l p_k = \dfrac{h}{2\pi i} \delta_{kl} \\ p_k p_l - p_l p_k = 0 \\ q_k q_l - q_l q_k = 0 \end{array}\right\} \quad (3)$$

Whence follows, as before, for any arbitrary function $f(q_1 \cdots q_f, p_1, \cdots p_f)$,

$$\left.\begin{array}{l} p_k f - f p_k = \dfrac{h}{2\pi i} \dfrac{\partial f}{\partial q_k} \\ f q_k - q_k f = \dfrac{h}{2\pi i} \dfrac{\partial f}{\partial p_k} \end{array}\right\} \quad (4)$$

Therefore the proof of the conservation of energy and of the frequency condition remains the same as above, as does the concept of canonic transformations

$$p_k = S p_k^0 S^{-1}, \qquad q_k = S q_k^0 S^{-1} \quad (5)$$

and the Hamilton-Jacobi equation,

$$H(pq) = S H(p_0 q_0) S^{-1} = W. \quad (6)$$

The large number of commutation rules gives rise to the question whether p_k, q_k can be at all determined so as to satisfy all the conditions. It is easily seen that all the condition equations are not independent. From the canonical equations of motion alone follows, for instance,

$$\frac{d}{dt} \sum_k (p_k q_k - q_k p_k) = 0.$$

The general proof of the possibility of satisfying all the conditions can be given by means of the theory of perturbations, starting from an unperturbed system with the energy-function

$$H_0(pq) = \sum_{k=1}^{f} H^{(k)}(p_k q_k)$$

which therefore consists of f uncoupled systems. Let the motions of these be represented by two-dimensional matrices

96   PROBLEMS OF ATOMIC DYNAMICS

$q_k{}^0$, $p_k{}^0$. If these f uncoupled systems are considered formally as a single system of f degrees of freedom, $q_k{}^0$, $p_k{}^0$ are to be represented by $2f$-dimensional matrices for which the relations

$$\left.\begin{array}{l} q_k{}^0(n_1 \cdots n_f, m_1 \cdots m_f) = \delta_k q_k{}^0(n_k m_k) \\ p_k{}^0(n_1 \cdots n_f, m_1 \cdots m_f) = \delta_k q_k{}^0(n_k m_k) \end{array}\right\}$$

hold where

$$\left.\begin{array}{l} \delta_k = 1 \text{ if } n_j = m_j \text{ for every } j \text{ except } j = k \\ \delta_k = 0 \text{ if for any } j \neq k,\, n_j \neq m_j. \end{array}\right\}$$

From these there follow in the first place the relations

$$\left.\begin{array}{l} p_k{}^0 q_l{}^0 - q_l{}^0 p_k{}^0 = 0 \quad \text{for } l \neq k \\ p_k{}^0 p_l{}^0 - p_l{}^0 p_k{}^0 = 0 \\ q_k{}^0 q_l{}^0 - q_l{}^0 q_k{}^0 = 0 \end{array}\right\} \tag{7}$$

Next the relation originally postulated for the 2-dimensional matrices,

$$p_k{}^0 q_k{}^0 - q_k{}^0 p_k{}^0 = \frac{h}{2\pi i}. \tag{8}$$

is also correct for $2f$-dimensional matrices.

If now the Hamiltonian function for the coupled system is

$$H = H_0 + \lambda H_1 + \lambda^2 H_2 + \cdots, \tag{9}$$

then we have shown that a solution of the unperturbed system exists which satisfies all the commutation relations. If we further suppose that the system H_0 is not degenerate, that is, that in the diagonal matrix W^0, which results from H_0 by the introduction of $q_k{}^0$, $p_k{}^0$, no two diagonal elements are equal, then we can find the motion of the perturbed system by the method of successive approximations discussed above. We place

$$q_k = S q_k{}^0 S^{-1}, \quad p_k = S p_k{}^0 S^{-1}$$
$$S = 1 + \lambda S_1 + \lambda^2 S_2 + \cdots$$

where S is determined by the equation

$$S H(p^0 q^0) S^{-1} = W.$$

THE STRUCTURE OF THE ATOM   97

The commutation relations and the equations of motion are then evidently satisfied and the desired proof is thus now complete.

The commutation relations are also invariant with respect to linear orthogonal transformations of q_k and p_k. For if we place

$$q_k' = \sum_l a_{kl} q_l$$
$$p_k' = \sum_l a_{kl} p_l \qquad \sum_l a_{kl} a_{jl} = \delta_{kj} \qquad (10)$$

Then

$$p_k' q_l' - q_l' p_k' = \sum_{ij} a_{ki} a_{lj}(p_i q_j - q_j p_i) = \frac{h}{2\pi i} \sum_j a_{kj} a_{lj} = \frac{h}{2\pi i} \delta_{kl}$$

and similarly for the other relations. Therefore, if we postulate that our fundamental relations hold for one cartesian system, they hold for any such system.

Proceeding systematically, we have now to study degenerate systems, that is, systems such that several of the W_n-values are equal, therefore several of the frequencies $\nu(nm)$ are zero. The constancy of the energy $\dot{H} = 0$ can still be deduced from the equations of motion and the commutation relations, but it *no longer* follows in general from $\dot{H} = 0$ that H is a diagonal matrix and therefore the proof of the frequency theorem cannot be carried through. The equations of motion and the commutation relations alone are here insufficient for a unique determination of the properties of the system and a further restriction of the fundamental equations is necessary. It is obvious that this restriction must be as follows: *As fundamental equations, the commutation relations and*

$$H = W = diagonal\ matrix \qquad (11)$$

shall hold. Then the validity of the frequency condition is also assured for degenerate systems.

Although, except in singular cases, the energy is uniquely determined by these conditions, the coördinates q_k are *not* uniquely determined. In non-degenerate systems, as was seen already in the example of the harmonic oscillator, only certain phase constants are arbitrary; one for each stationary state. In degenerate systems a much greater indetermination exists, which evidently is related to a sort of lability which allows arbitrarily small external perturbations to produce finite changes in the coördinates. But it can be shown that even then those properties of the system on which the polarization of the emitted light depends vary only continuously, a fact which Heisenberg has called "spectroscopic stability." We shall not discuss this question further.

LECTURE 16

Conservation of angular momentum — Axial symmetrical systems and the quantization of the axial component of angular momentum.

The applications of the basic principles, which have been considered so far, suppose that several specially simple systems which are used as starting points in the calculus of perturbations are completely known. For this purpose we have until now studied the particular example of the harmonic oscillator. We must now develop general methods for the direct integration of the fundamental equations. These methods are the same as those used in classical mechanics, i.e., general properties of the energy-function H are used to find integrals. The conservation of energy has been thus derived as a consequence of the property of H not to depend explicitly on the time. The conservation of momentum and moment of momentum will now be developed, making the same hypotheses on H as in ordinary mechanics. The integration method is quite similar to that used in the derivation of the conservation of energy. The equations of motion, considered for the elements of the matrices, form an infinite system for an infinite number of variables: in general each equation contains an infinite number of variables. To begin with, a function $A(pq)$, constant according to the fundamental equations, and therefore a diagonal matrix for non-degenerate systems is determined. If $\phi(pq)$ is any function, the difference

$$\phi A - A\phi = \psi$$

can be calculated from our commutation rules. But as A is a diagonal matrix each of the equations for the elements contains only one of the elements of ϕ and ψ, besides two diagonal terms of A.

In Galilean-Newtonian mechanics, as well as in Einstein's ("relativistic") mechanics:

$$H = H'(p) + H''(q). \tag{1}$$

The components of momentum are

$$p_x = \sum_{k=1}^{3} p_{kx}$$
$$p_y = \sum_{k=1}^{3} p_{ky} \qquad (2)$$
$$p_z = \sum_{k=1}^{3} p_{kz}$$

and the components of moment of momentum

$$M_x = \sum_{k=1}^{3} (q_{ky}p_{kz} - p_{ky}q_{kz})$$
$$M_y = \sum_{k=1}^{3} (q_{kz}p_{kx} - p_{kz}q_{kx}) \qquad (3)$$
$$M_z = \sum_{k=1}^{3} (q_{kx}p_{ky} - p_{kx}q_{ky})$$

If derivatives with respect to time are now taken and it is noted that, because of our assumption on H, all $\dot{p}_{kx}\cdots$ etc. depend only on $q_{kx}\cdots$ etc., and all $\dot{q}_{kx}\cdots$ etc. on $p_{kx}\cdots$ etc., it is seen that all these derivatives have the form $\phi(q) + \psi(p)$. Now since all the q's and all the p's are interchangeable among themselves, these expressions all vanish under the same conditions as in classical mechanics. The theorems on uniform motion of the center of gravity and on the conservation of angular momentum (law of areas) therefore hold exactly as in classical theory.

Let us now build up the expression

$$M_x M_y - M_y M_x = \sum_{k,l} \Big[(q_{ky}p_{kz} - p_{ky}q_{kz})(q_{lz}p_{lx} - p_{lz}q_{lx})$$
$$- (q_{kz}p_{kx} - p_{kz}q_{kx})(q_{ly}p_{lz} - p_{ly}q_{lz}) \Big]$$
$$= \sum_{k,l} \Big[q_{ky}p_{lx}(p_{kz}q_{lz} - q_{lz}p_{kz}) + p_{ky}q_{lx}(q_{kz}p_{lz} - p_{lz}q_{lz}) \Big]$$
$$= -\frac{h}{2\pi i} \sum_{k} (q_{kx}p_{ky} - p_{kx}q_{ky})$$

therefore

$$M_x M_y - M_y M_x = -M_z \epsilon, \; \epsilon = \frac{h}{2\pi i}. \qquad (4)$$

whence it is seen that the law of areas, as in classical mechanics, holds either only for one or for all three axes.

It will now be assumed that the system consists only of discrete energy levels, further that it is not degenerate and that the law of areas holds for one of the momenta, for instance $\dot{M}_z = 0$. This is e.g. the case if the external forces acting on the atom are symmetrical with respect to the z-axis. Then M_z is a diagonal matrix and the individual elements M_{zn} are to be interpreted as the angular momenta of the atom around the z-axis for the corresponding individual states.

From the definition of M_x, M_y, M_z and the commutation rules follow the matrix equations

$$\left. \begin{array}{l} q_{lx} M_z - M_z q_{lx} = +\epsilon q_{ly} \\ q_{ly} M_z - M_z q_{ly} = -\epsilon q_{lx} \\ q_{lz} M_z - M_z q_{lz} = 0 \end{array} \right\} \qquad (5)$$

As

$$M_z(nm) = \delta_{nm} M_{zn},$$

these expressions can be rewritten

$$\left. \begin{array}{l} q_{lx}(nm)(M_{zn} - M_{zm}) = +\epsilon q_{ly}(nm) \\ q_{ly}(nm)(M_{zn} - M_{zm}) = -\epsilon q_{lx}(nm) \\ q_{lz}(nm)(M_{zn} - M_{zm}) = 0 \end{array} \right\} \qquad (6)$$

Equations (6) express, in the ordinary language of Bohr's theory, the following: For a quantum jump in which the angular momentum M_{zn} changes, $q_{lz}(nm) = 0$ and the plane of oscillation of the emitted light wave is therefore perpendicular to the z-axis. For jumps in which M_{zn} does not change $q_{lx}(nm) = 0$, $q_{ly}(nm) = 0$ and the emitted light therefore vibrates parallel to the z-axis. Moreover, in the former case,

$$\left[(M_{zn} - M_{zm})^2 - \frac{h^2}{4\pi^2}\right] q_{l\eta}(nm) = 0, \; \eta = x, y. \qquad (7)$$

102 PROBLEMS OF ATOMIC DYNAMICS

That is: for every quantum jump M_{zn} changes by 0 or by $\pm \dfrac{h}{2\pi}$. In the former case the emitted light is linearly polarized parallel to the z-axis, in the latter case it is circularly polarized around this axis. M_{zn} can therefore be represented by

$$M_{zn} = \frac{h}{2\pi}(n_1 + C), \qquad n_1 = \cdots -2, -1, 0, 1, 2 \cdots. \quad (8)$$

If states existed the angular momentum of which did not find a place in this series, there could be no jumps or interactions between them and those belonging to the above series.

From these results it is seen that the index n can be split up into two components, one of which is the number n_1 which has already been introduced, while the other, n_2, numbers the different n's with the same n_1. Our matrices become four-dimensional and the "polarization rules" already derived are equivalent to the following expressions:

$$\left.\begin{array}{r}q_{lx}(nm) = \delta_{1,|n_1-m_1|}q_{lx}(nm) \\ q_{ly}(nm) = \delta_{1,|n_1-m_1|}q_{ly}(nm) \\ q_{lz}(nm) = \delta_{n_1,m_1}q_{lz}(nm) \\ q_{lx}(n_1, n_2; n_1 \pm 1, m_2) \mp iq_{ly}(n_1, n_2; n_1 \pm 1, m_2) = 0\end{array}\right\} \quad (9)$$

All these relations hold if q_{lx}, q_{ly}, q_{lz} are replaced by p_{lx}, p_{ly}, p_{lz} or by M_x, M_y, M_z.

In particular we note that

$$\left.\begin{array}{r}M_x(nm) = \delta_{1,|n_1-m_1|}M_x(nm) \\ M_y(nm) = \delta_{1,|n_1-m_1|}M_y(nm) \\ M_x(n_1, n_2; n_1 \pm 1, m_2) \mp iM_z(n_1, n_2; n_1 \pm 1, m_2) = 0\end{array}\right\}$$

Further we need the following derived commutation relations: If,

$$\mathbf{q}_l^2 = q_l^2 = q_{lx}^2 + q_{ly}^2 + q_{lz}^2, \qquad \mathbf{M}^2 = M^2 = M_x^2 + M_y^2 + M_z^2,$$

then simple calculations give

$$\left.\begin{array}{r}q_l^2 M_z - M_z q_l^2 = 0 \\ M^2 M_z - M_z M^2 = 0\end{array}\right\} \quad (10)$$

which means that q_l^2 and M^2 are diagonal matrices with respect to the quantum number n_1.

The two components M_x, M_y may also be constant, but can never be diagonal matrices. For from

$$M_y M_z - M_z M_y = -\epsilon M_x$$

or

$$M_y(nm)(M_{zn} - M_{zm}) = -\epsilon M_x(nm)$$

it would follow that, for $M_y(nm) = M_{yn}\delta_{nm}$, M_x would vanish identically, and therefore that M_y, M_z would also vanish identically. Such a system with a constant vector **M**, for instance a system moving freely in space, is hence necessarily degenerate.

Consider now a system the energy function of which is

$$H = H_0 + \lambda H_1 + \cdots$$

under the following assumptions: *For $\lambda = 0$ the law of areas holds for all three directions. For $\lambda \neq 0$ the system is not degenerate, but M_z is constant. The energy H_0 does not depend on n_1.* A system of this kind is, for instance, an atom in an axially symmetrical field of strength proportional to λ. This investigation leads also to definite information about the degenerate system with the energy-function H_0, for every property of the perturbed system which is independent of λ or the choice of the privileged direction z must remain valid for $\lambda = 0$.

According to our hypothesis that for $\lambda = 0$ the law of areas along all three directions holds, $\dot{M}_x$, $\dot{M}_y$ and therefore also $\dfrac{d}{dt}(M^2)$ have no terms without λ. Therefore

$$\begin{aligned}\nu_0(nm)M_x{}^0(nm) &= 0 \\ \nu_0(nm)M_y{}^0(nm) &= 0 \\ \nu_0(nm)(M^0)^2(nm) &= 0.\end{aligned} \quad (11)$$

As it has been further supposed that $H_0 = W^0$ is independent of the quantum number n_1, we have

$$\nu_0(n_1, n_2; m_1, n_2) = W_{n_2}{}^0 - W_{n_2}{}^0 = 0$$

$$\nu_0(n_1, n_2; m_1, m_2) = W_{m_2}{}^0 - W_{m_2}{}^0 \neq 0 \text{ for } n_2 \neq m_2,$$

whence

$$M_x{}^0(nm) = \delta_{n_2m_2}M_x{}^0(nm)$$
$$M_y{}^0(nm) = \delta_{n_2m_2}M_y{}^0(nm) \qquad (12)$$
$$(M^0)^2(nm) = \delta_{n_2m_2}(M^0)^2(nm)$$

It was shown earlier that M^2 is quite in general a diagonal matrix with respect to n_1; it is now shown that $(M^0)^2$ is a diagonal matrix with respect to both quantum numbers n_1, n_2. The same holds for $(M_x{}^0)^2 + (M_y{}^0)^2 = (M^0)^2 - M_z{}^2$. Now

$$(M_x{}^0)^2(n_1n_2m_1m_2) = \sum_{k_1k_2} M_x{}^0(n_1n_2k_1k_2)M_x{}^0(k_1k_2m_1m_2)$$
$$= \delta_{n_2m_2}\sum_{k_1} M_x{}^0(n_1n_2k_1n_2)M_x{}^0(k_1n_2m_1n_2)$$

and

$$((M_x{}^0)^2 + (M_y{}^0)^2)(n_1n_2m_1n_2)$$
$$= \delta_{n_1m_1}\delta_{n_2m_2}\sum_{k_1}\{M_x{}^0(n_1n_2k_1n_2)M_x{}^0(k_1n_2n_1n_2)$$
$$+ M_y{}^0(n_1n_2k_1n_2)M_y{}^0(k_1n_2n_1n_2)\}$$
$$= \delta_{n_1m_1}\delta_{n_2m_2}\sum_{k_1}[\,|M_x{}^0(n_1n_2k_1n_2)|^2 + |M_y{}^0(n_1n_2k_1n_2)|^2] \qquad (13)$$

The diagonal terms of $(M^0)^2 - M_z{}^2$ are therefore always positive; and since $(M^0)^2$ does not depend on n_1, the number of possible values of $M_{zn_1}^2 = \left(\dfrac{h}{2\pi}\right)^2(n_1 + C)^2$ for a given n_2, therefore for a given $(M_{n_2}^0)^2$, is finite. In other words the number of values n_1 for a given n_2 is finite. Hence the sum

$$\sum_{k_1k_2} M_x{}^0(n_1, n_2;\ k_1, k_2)M_y{}^0(k_1, k_2;\ m_1, m_2)$$
$$= \delta_{n_2m_2}\sum_{k_1} M_x{}^0(n_1, n_2;\ k_1, n_2)M_y{}^0(k_1, n_2;\ m_1, n_2)$$

has only a finite number of terms. This sum is an element of $M_x{}^0M_y{}^0$. If we now form in the same way $M_y{}^0M_x{}^0$ and sum the equations

$$-\epsilon M_z{}^0 = M_x{}^0M_y{}^0 - M_y{}^0M_x{}^0$$

THE STRUCTURE OF THE ATOM

over n_1 for fixed n_2, this sum is zero on the right-hand side because in general, for *finite* matrices, the diagonal sum of ab is equal to that of ba:

$$\sum_n (\sum_k a(nk)b(kn)) = \sum_n (\sum_k b(nk)a(kn)).$$

Therefore

$$\sum_{n_1} M_z = \frac{h}{2\pi} \sum_{n_1} (n_1 + C) = 0. \tag{14}$$

This holds for every complete series of n_1. Therefore the possible values of n_1 which go with a fixed n_2 always form a symmetrical series with respect to the origin. Hence $(n_1 + C)$ runs through a finite series of whole numbers $\cdots -2, -1, 0, 1, 2 \cdots$ or of "half-numbers" $\cdots -\frac{3}{2}, -\frac{1}{2}, +\frac{1}{2}, +\frac{3}{2}, +, \cdots$.

In the literature m (magnetic quantum number) is used in place of $n_1 + C$. It has therefore been shown that the quantum number m defined by the diagonal term of $M_z = \frac{h}{2\pi} m$ is either a whole or a half number and that the selection rule

$$m \rightarrow \begin{cases} m + 1 \\ m \\ m - 1 \end{cases} \tag{15}$$

holds.

This result does not seem to lead much further than that which was obtained from the classical theory of multiple-periodic systems, but it must be borne in mind that in classical theory certain orbits frequently had to be ruled out by additional excluding rules. For instance, in the theory of the hydrogen atom, orbits leading to a collision between the electron and the nucleus were excluded. In the present theory no such additional rules are necessary, a fact which must be regarded as an essential step forward. To this must be added the full justification of half and whole quantum numbers, which so far could not be explained theoretically, while the empirical facts necessarily led to the introduction of the former, as already shown.

LECTURE 17

Free systems as limiting cases of axially symmetrical systems — Quantization of the total angular momentum — Comparison with the theory of directional quantization — Intensities of the Zeeman components of a spectral line — Remarks on the theory of Zeeman separation.

The detailed presentation of the derivations in the preceding lecture are, I believe, sufficient to show the method clearly. From now on I shall mainly emphasize results. Proceeding along the same line of reasoning we arrive at a new quantum number j which determines, in the limit $\lambda \to 0$, the diagonal terms of M^2, as follows,

$$\text{Diagonal terms of } M^2 = \left(\frac{h}{2\pi}\right)^2 j(j+1). \tag{1}$$

Further j is always equal to the maximum value of the quantum number m and therefore is a whole or a half number. The selection rule

$$j \to \begin{cases} j+1 \\ j \\ j-1 \end{cases} \tag{2}$$

holds. The proof is quite similar to that in classical theory. In the latter a new rectangular system of coördinates is introduced whose z-axis coincides, for $\lambda = 0$, with the fixed direction of the angular momentum. Considerations concerning the total angular momentum are quite similar, for this system, to those for M_z in the case of axial symmetry. A linear combination of the coördinate matrices is formed which corresponds formally to a rotation of the system of coördinates into the desired position (z-axis parallel to the angular momentum). The equations obtained from these expressions have a finite number of matrix elements of a type similar to that obtained previously for the coördinates themselves except that M^2 occurs

THE STRUCTURE OF THE ATOM

instead of M_z. We find from M_z and M by means of the identity

$$(M_x + iM_y)(M_x - iM_y) = M_x^2 + M_y^2 - i(M_xM_y - M_yM_x)$$
$$= M^2 - M_z^2 + i\epsilon M_z$$

and the above relations for M_x and M_y [Equations (3), (4), Lecture (16)]

$$\left.\begin{array}{l}(M_x+iM_y)(j, m-1; j, m) = \dfrac{1}{2}\dfrac{h}{2\pi}\sqrt{j(j+1)-m(m-1)} \\[2mm] (M_x-iM_y)(j, m; j, m-1) = \dfrac{1}{2}\dfrac{h}{2\pi}\sqrt{j(j+1)-m(m-1)}\end{array}\right\} \quad (3)$$

It is also possible to express explicitly the coördinates q_{lx}, q_{ly}, q_{lz} in terms of the quantum numbers m, j. The result can be most clearly stated if written separately for the three possible jumps of j.

$$j \to j \left\{\begin{array}{l}(q_{lx}+iq_{ly})(j, m-1; j, m) = A\tfrac{1}{2}\sqrt{j(j+1)-m(m-1)} \\[2mm] (q_{lx}-iq_{ly})(j, m; j, m-1) = A\tfrac{1}{2}\sqrt{j(j+1)-m(m-1)} \\[2mm] q_{lz}(j, m) = Am\end{array}\right. \quad (4)$$

$$j \to j-1 \left\{\begin{array}{l}(q_{lx}+iq_{ly})(j, m-1; j-1, m) = B\tfrac{1}{2}\sqrt{(j-m)(j-m+1)} \\[2mm] (q_{lx}-iq_{ly})(j, m; j-1, m-1) = -B\tfrac{1}{2}\sqrt{(j+m)(j+m-1)} \\[2mm] q_{lz}(j, m; j-1, m) = B\sqrt{j^2-m^2}\end{array}\right. \quad (5)$$

$$j \to j+1 \left\{\begin{array}{l}(q_{lx}+iq_{ly})(j, m; j+1, m+1) = C\tfrac{1}{2}\sqrt{(j+m+2)(j+m+1)} \\[2mm] (q_{lx}-iq_{ly})(j, m; j+1, m-1) = C\tfrac{1}{2}\sqrt{(j-m+2)(j-m+1)} \\[2mm] q_{lz}(j, m; j+1, m) = C\sqrt{(j+1)^2-m^2}\end{array}\right. \quad (6)$$

where A, B, C, depend in some way on the other quantum numbers of the system.

These expressions, as the perturbation and dispersion formulas given above (Lecture 14), had been found previously by considerations of correspondence before they were derived by the methods of our theory. This is seen most easily from formula (4) for $j \to j$ by going over to the limiting case of large quantum numbers m, j. Then 1 can be neglected compared with m and j and we find for the ratio of the intensities of the two circular vibrations and the linear oscillation:

$$|q_{lx}+iq_{ly}|^2 : |q_{lx}-iq_{ly}|^2 : |q_{lz}|^2 = \tfrac{1}{4}(j^2-m^2) : \tfrac{1}{4}(j^2-m^2) : m^2$$
$$= \tfrac{1}{4}(M^2-M_z^2) : \tfrac{1}{4}(M^2-M_z^2) : M_z^2$$
$$= \tfrac{1}{4}(M_x^2+M_y^2) : \tfrac{1}{4}(M_x^2+M_y^2) : M_z^2 \quad (7)$$

where M, M_x, M_y, M_z denote the total angular momentum and its components in the quantized state in question m, j. These formulas can, however, be obtained classically as follows:

Consider the motion of the electrons as represented by the motion of their electric center of gravity S. The motion of S is decomposed into a linear component parallel to the angular momentum **M** and two circular components rotating in opposite directions perpendicular to **M**; the first component corresponds alone to the jump $j \to j$, the two others correspond to jumps $j \to j \pm 1$. The linear oscillation parallel to **M** is given by

$$\left. \begin{array}{l} q_x = a \sin \omega t \cos \phi \sin \theta \\ q_y = a \sin \omega t \sin \phi \sin \theta \\ q_z = a \sin \omega t \cos \theta \end{array} \right\}$$

where ϕ, θ are the polar coördinates of the direction of **M** in a fixed system of coördinates. The motion in the xy-plane can be decomposed into two rotations in opposite directions,

$$\left. \begin{array}{l} q_x = q_x' + q_x'' \\ q_y = q_y' + q_y'' \end{array} \right\}$$

where

$$q_x' = \frac{a}{2}\sin\theta \sin(\omega t + \phi), \quad q_x'' = \frac{a}{2}\sin\theta \sin(\omega t - \phi)$$

$$q_y' = -\frac{a}{2}\sin\theta \cos(\omega t + \phi), \quad q_y'' = \frac{a}{2}\sin\theta \cos(\omega t - \phi)$$

To these correspond the jumps $m \to m \pm 1$, while the component q_z corresponds to jumps $m \to m$. The intensities are proportional to

$$q_x'^2 + q_y'^2 : q_x''^2 + q_y''^2 : q_z^2 = \frac{a^2}{4}\sin^2\theta : \frac{a^2}{4}\sin^2\theta : a^2\cos^2\theta. \quad (8)$$

If a weak outer field is now established in the z-direction, the whole atom rotates slowly around this direction. The circular frequencies of the two circular component oscillations are thereby changed slightly as well as their intensities, but in the limiting case of an infinitely weak field these changes can be neglected. Then we have

$$M_z = M\cos\theta, \quad \sqrt{M_x^2 + M_y^2} = M\sin\theta.$$

If these are introduced in the above relations the formula (8) given above is shown to be the limiting value of the rigorous formula (4), obtained through the new quantum theory. The cases $j \to j \pm 1$ can be interpreted in a precisely similar manner.

Historically, in fact, the opposite has been done. Starting from the classical motion, intensity formulas have been found, which should be correct for large quantum numbers, but which need a correction for smaller m and j. This correction has been found in a number of ways. Goudsmit and Kronig have used the so-called "boundary principle," that is the requirement that the intensities must vanish when either of the states disappears. It was noted above that j is the maximum value of m; therefore, the intensities of all jumps for which j remains unchanged but m changes vanish if we place $m = j + 1$. Our formula is seen to satisfy this condition.

The first incentive to the investigation of these intensity

110   PROBLEMS OF ATOMIC DYNAMICS

laws was the experimental researches on the relative brightness of the components of the Zeeman effect. These, under the direction of Ornstein, were carried out by Moll, Burgers, Dorgelo and others. These investigators at Utrecht first found empirical whole-number laws for the intensities in the Zeeman effect and gave simple rules for their calculation. The theory then developed as outlined above.

Our formulas fit exactly the case of an atom in a weak magnetic field. The number and position of the components into which a line is split up by the field cannot as yet be calculated theoretically. We shall return to this question later. But, if we consider the system of split-up lines given experimentally, we can read off from it the value of j. Thus, starting from the middle, the Zeeman components vibrating parallel to the field ($m \to m$) are assigned half or whole numbers and the largest value of m is equal to j. We have therefrom the corresponding values of m and j for each line and hence can calculate the relative intensities by our formula. Comparisons with observations have in all cases verified the theory.

As already stated, the actual magnetic split of spectral lines is not given by the present theory as it stands. For if linear terms in the field strength, such as

$$\frac{e}{2\mu c} |\mathbf{H}| M_z$$

are alone taken into account in the magnetic additional terms of the energy, as usual, then we obtain, on account of our formula for M_z, the equidistant term sequence

$$\frac{e}{4\pi\mu c} |\mathbf{H}| m = \nu_m m$$

with the *normal* separation ν_m corresponding to the classical Larmor precession. Experimentally, however, the separation is found to be $g\nu_m$; the numbers g have been determined empirically by Landé as functions of the quantum numbers characterizing the corresponding spectral lines. All attempts to derive these g-formulas from classical models led to similar formulas, but never the correct ones. In these formulas the square of the angular momentum M^2 enters, among other quantities. The former, of course, has always been replaced so far by $(jh/2\pi)^2$, but Landé's empirical formulas always require the expression $j(j+1)$ instead of j^2. Our new quantum theory gives in fact $M^2 = \left(\dfrac{h}{2\pi}\right)^2 j(j+1)$, a circumstance which encourages us to further researches.

We have now to turn our attention to a new point which distinguishes the new from the classical theory and may result in rendering Larmor's theorem *invalid:* This new point is the neglect of the terms in $\mathbf{H}^2$ entering in the expression for the magnetic energy. These can certainly be neglected in the classical theory for orbits of small dimensions but can *not* be neglected for orbits of large dimensions or hyperbolic paths. In the limiting case of a free electron the period of revolution is just twice the normal Larmor precession. In quantum mechanics all these orbits, the distant as well as the near ones, are so intimately connected due to the peculiar kinematics and geometry that the justification of neglecting $\mathbf{H}^2$ is no longer evident, for the probability of transition even from the unexcited state to that of a free electron is always considerable. For the oscillator we have certainly the normal Zeeman effect. For the nuclear atom, however, it is possible that the intimate connection between inner and outer orbits may lead to different results. There are, on the other hand, powerful arguments against such an explanation, particularly the intimate connection between the anomalous Zeeman effect and the multiplet structure of spectral lines. A new physical concept seems to be required here. Such an idea has been formulated by Uhlenbeck and Goudsmit, but here I can only indicate it. Pauli, from the study of multiplets, has been led to attribute to each electron not three quantum numbers, as would correspond to its number of degrees of freedom, but *four.* Until now this has been considered as something purely formal, to be eliminated if possible. Uhlenbeck and Goudsmit, however, take this hypothesis earnestly. They attribute to the electron a proper rotation and a corresponding magnetic field determined by the fourth quantum number. Preliminary calculations by Heisenberg and Jordan have shown that this idea forms a basis for an exact theory of the abnormal Zeeman effect, but I am unable to give further details on this at present.

LECTURE 18

Pauli's theory of the hydrogen atom.

We now come to the crucial question for the whole new theory: Is it able to account for the properties of the hydrogen atom? Let us recall that the explanation of the hydrogen spectrum (Balmer's formula) was the first great success of Bohr's theory and has since remained its keynote. If the new theory failed here it would have to be abandoned in spite of its many conceptual advantages, but, as Pauli has shown, it stands the test successfully.[*] I can give here only the fundamental ideas and results of this development, as yet unpublished.

In the classical theory of Keplerian motion it is customary to operate with polar coördinates. This process fails here because it does not seem possible to consider angular variables as matrices. Pauli avoided this difficulty by retaining rectangular coördinates and introducing an additional coördinate, the radius vector r, which is related to x, y, z, by the relation

$$r^2 = x^2 + y^2 + z^2.$$

The process will first be explained using the classical model. We have the energy function

$$H = \frac{1}{2\mu}\mathbf{p}^2 - \frac{Ze^2}{r} \qquad (1)$$

and the equations of motion,

$$\dot{\mathbf{r}} = \frac{1}{\mu}\mathbf{p}, \qquad \dot{\mathbf{p}} = -\frac{Ze^2\mathbf{r}}{r^3}. \qquad (2)$$

From these follows that the angular momentum

$$\mathbf{M} = \mathbf{r} \times \mathbf{p} \qquad (3)$$

is constant with respect to time. There follows, further, by using the relation

$$\mathbf{M} \times \mathbf{r} = (\mathbf{r} \times \mathbf{p}) \times \mathbf{r} = \mathbf{p}r^2 - (\mathbf{p} \cdot \mathbf{r})\mathbf{r}$$

that the vector

$$\mathbf{A} = \frac{1}{Ze^2\mu}\mathbf{M} \times \mathbf{p} + \frac{\mathbf{r}}{r} \qquad (4)$$

is also constant with respect to time. Placing $M = |\mathbf{M}|$ we obtain at once

$$\mathbf{A} \cdot \mathbf{r} = -\frac{1}{Ze^2\mu}M^2 + r,$$

which is the equation of a conic. If we take the xy-plane in the plane of this curve and the x-axis in the direction of $\mathbf{A}$, that is

$$\begin{aligned} x &= r\cos\phi & A_x &= |\mathbf{A}| = A \\ y &= r\sin\phi & A_y &= 0 \\ z &= 0 & A_z &= 0 \end{aligned}$$

we obtain

$$Ar\cos\phi = -\frac{M^2}{Ze^2\mu} + r$$

or

$$r = \frac{M^2}{Ze^2\mu}\left(\frac{1}{1 - A\cos\phi}\right).$$

A is therefore the eccentricity and we find for the energy

$$W\frac{2}{Z^2e^4\mu}M^2 = A^2 - 1. \qquad (5)$$

This calculation can be repeated, with only slight changes, in matrix mechanics.

The matrices x, y, z, r are commutative among themselves, as also the momentum matrices p_x, p_y, p_z, p_r. The following are also commutative:

$$x \text{ with } p_y, p_z \cdots$$
$$p_x \text{ with } y, z \cdots$$

but,

$$p_x x - x p_x = \frac{h}{2\pi i}, \text{ etc.}$$

THE STRUCTURE OF THE ATOM

The energy and the equations of motion are the same as above. From the latter there follows at once, as has been shown generally before, that the angular momentum is constant in time.

It can be shown further that the vector

$$\mathbf{A} = \frac{1}{Ze^2\mu} \tfrac{1}{2}(\mathbf{M} \times \mathbf{p} + \mathbf{p} \times \mathbf{M}) + \frac{\mathbf{r}}{r}$$

is constant in time. To prove this a longer calculation is necessary and secondary commutation relations are needed, as for instance

$$yM_z - M_z y = -\frac{h}{2\pi i}x, \qquad M_y z - zM_y = \frac{h}{2\pi i}$$

$$\dots\dots\dots\dots\dots\dots\dots\dots\dots\dots\dots\dots\dots\dots\dots\dots$$

$$p_x r - rp_x = \frac{h}{2\pi i}\frac{x}{r}, \dots.$$

$$p_x \frac{x}{r} - \frac{x}{r}p_x = \frac{h}{2\pi i}\frac{y^2 + z^2}{r^3}, \dots p_x \frac{y}{r} - \frac{y}{r}p_x = -\frac{h}{2\pi i}\frac{xy}{r^3}.$$

Derivatives with respect to time are transformed by means of the formula

$$\frac{d}{dt}\frac{x}{r} = \frac{2\pi i}{h}\left(W\frac{x}{r} - \frac{x}{r}W\right) = \frac{2\pi i}{h}\frac{1}{2\mu}\left(p^2\frac{x}{r} - \frac{x}{r}p^2\right).$$

The problem is now to find the constant vectors **M** and **A**. For these the following commutation relations hold:

$$\left.\begin{aligned} M_x M_y - M_y M_x &= -\frac{h}{2\pi i}M_z \cdots \\ A_x M_y - M_y A_x &= -\frac{h}{2\pi i}A_z \cdots \\ M_x A_y - A_y M_x &= -\frac{h}{2\pi i}A_z \cdots \end{aligned}\right\} \quad (6)$$

$$A_x A_y - A_y A_x = \frac{h}{2\pi i}\frac{2}{\mu Z^2 e^4} W M_z \cdots \quad (7)$$

Finally, the following equation is found

$$\left(M^2 + \frac{h^2}{4\pi^2}\right)\frac{2}{\mu Z^2 e^4} W = A^2 - 1. \quad (8)$$

This equation differs from the corresponding classical equation only by the term $\dfrac{h^2}{4\pi^2}$ added to M^2. This is just one important characteristic of the new theory.

In the solution of these equations W is always a diagonal matrix, but the constant components of the vector matrices **p**, **A** are not diagonal matrices, as was shown in general above. According to our previous results, the requirement that, besides W, p_z and p^2 be diagonal matrices has a definite meaning, i.e., the addition of a weak axially symmetrical perturbing field of force the energy of which depends on p_z and p.

The same method used above (Lecture 16) can now be applied to determine the vector **P**. Then Equations (6) are exactly the same as Equations (4) and (5), Lecture 16, except that the coördinates q_{lx}, q_{ly}, q_{lz} are replaced by A_x, A_y, A_z. Instead of the quantum numbers n_1, n_2 we write the usual symbols k and m, where k determines the total angular momentum, denoted by j above, and m its z-component. We then have

$$p_z(k, m; k, m) = \frac{h}{2\pi} m, \qquad p^2 = \frac{h^2}{4\pi^2} k(k+1)$$

$$|p_x(k, m; k, m \pm 1)|^2 = |p_y(k, m; k, m \pm 1)|^2$$

$$= \frac{1}{4}\frac{h^2}{4\pi^2}[k(k+1) - m(m+1)] \qquad (9)$$

where m runs through a complete series of half or whole numbers from $-k$ to $+k$. Further we obtain for A_x, A_y, A_z expressions quite similar to those found before for q_{lx}, q_{ly}, q_{lz}, for example the following one:

$$|A_x(k+1, m; k, m \pm 1)|^2 = |A_y(k+1, m; k, m \pm 1)|^2$$

$$= \tfrac{1}{4}C(k+1, k)(k \mp m)(k \mp m + 1)$$

$$|A_z(k+1, m; k, m)|^2 = C(k+1, k)((k+1)^2 - m^2).$$

Consider in Equations (7) a given W and the smallest possible k. Then a closer discussion shows that the equations in question can be satisfied if m is zero and only zero, hence $k_{\min} = 0$. Herein is contained the *integrality* of k and m.

THE STRUCTURE OF THE ATOM

Formula (7) gives further for the functions $C(k+1, k) = C(k, k+1)$ the equation,

$$(2k-1)C(k, k-1) - (2k+3)C(k+1, k) = \frac{|W|}{Rh}, k \neq 0.$$

W is here assumed to be *negative*, i.e., ellipse, not hyperbola; R is Rydberg's constant. As solution of this equation we obtain

$$C(k+1, k) = \frac{|W|}{kh} \frac{(k_m - k)(k_m + k + 2)}{(2k+1)(2k+3)}$$

where k_m is the maximum value of k for a given $|W|$. We now have the components of **A** and, therefore, also the value of

$$A^2 = A_x^2 + A_y^2 + A_z^2,$$

that is,

$$\begin{aligned} A^2(k, m; k, m) &= (k+1)(2k+3)C(k+1, k) \\ &\quad + k(2k-1)C(k, k-1) \\ &= \frac{|W|}{Rh}[k_m^2 + 2k_m - k(k+1)]. \end{aligned} \quad (10)$$

There follows finally from Equation (8):

$$1 = \frac{|W|}{Rh}(k_m + 1)^2.$$

If we write $n = k_m + 1$, then n corresponds to the main quantum number of Bohr's theory and takes the values 1, 2, 3, $\cdots$. For a given n, k has the values $k = 0, 1, 2, \cdots n - 1$. We have thus found Balmer's formula

$$W = -\frac{Rh}{n^2} \quad (n = 1, 2, 3, \cdots) \quad (11)$$

and have shown at the same time how each term is split up on removal of the degeneration by the addition of weak perturbing forces. This split is given by $k = 0, 1, 2, \cdots n - 1$; $m = -k, -k + 1, \cdots k - 1, k$.

A characteristic trait of the new theory is that the value $k = n$ does not occur. It follows in particular that in the unexcited state $n = 1$, $k = 0$ and, therefore, $m = 0$; in other words, the normal state is not magnetic. This result must be revised, however, if the rotating magnetic electron of Uhlenbeck and Goudsmit is accepted.

Pauli has succeeded in deriving in a similar way the Stark effect for the hydrogen atom. In this case also no additional conditions need be imposed. The same holds in the case where an electric and a magnetic field act in arbitrary directions (crossed fields). Just here the classical theory of multiple-periodic systems encountered great difficulties, for the frequencies of the overtones are made up of two fundamental periods (the electric frequency ν_e and the magnetic Larmor frequency ν_m) and, therefore, protracted commensurabilities appear when the fields are varied, that is equations of the form

$$\tau_1 \nu_e + \tau_2 \nu_m = 0$$

with integral τ_1, τ_2. This means that arbitrarily small adiabatic changes, of the electric field, for instance, will produce degeneration. The validity of Ehrenfest's adiabatic hypothesis is no longer certain and, therefore, the quantum rules become doubtful. All these difficulties disappear in the new theory.

Pauli has also attacked the theory of fine structure (relativistic change of mass), without yet quite attaining his goal.

LECTURE 19

Connection with the theory of Hermitian forms — Aperiodic motions and continuous spectra.

Let us now inquire how aperiodic motions, such as hyperbolic orbits in the hydrogen atom, can be treated in the new theory. It is to be expected *a priori* that there is no essential difference in the treatment of periodic and aperiodic processes because the postulate of periodicity does not appear explicitly in the fundamental equations. The notion of matrix can be generalized at once so as to permit the representation of aperiodic processes. The indices n, m have only to be considered as continuous variables and the matrix product defined by the integral

$$pq = \left(\int p(nk)q(km)dk \right),$$

but difficulties appear at once if we attempt to generalize the notion of unit matrix to include these continuous matrices. This must be done because the unit matrix enters in the commutation relation

$$pq - qp = \frac{h}{2\pi i} \cdot 1. \qquad (1)$$

That function $f(nm)$ is to be taken as unit matrix which vanishes for $n \neq m$ and becomes infinite for $n = m$ in such a way that the integrals

$$\int f(nk)dk \quad \text{and} \quad \int f(kn)dk$$

become unity, for then

$$qf = \left(\int q(nk)f(km)dk \right) = (q(nk)) = q$$

and at the same time $fq = q$. It is clear that operating with such unusual functions is not convenient. In circumventing

this difficulty a way indicated by entirely different lines of reasoning has been followed. In classical mechanics the known theory of oscillation of a system is intimately related to the theory of quadratic forms. Oscillations occur when the potential energy is a "definite" quadratic form of the variables, i.e., one which does not change its sign. For two variables x, y, for instance,
$$U = \tfrac{1}{2}(a_{11}x^2 + 2\,a_{12}xy + a_{22}y^2).$$
Oscillations are obtained in the simplest way by transforming this form to a sum of squares by means of the linear transformation,
$$x = h_{11}\xi + h_{12}\eta$$
$$y = h_{21}\xi + h_{22}\eta.$$
We now try to effect this transformation in such a way that the kinetic energy $T = \dfrac{m}{2}(\dot{x}^2 + \dot{y}^2)$, which is already a sum of squares, retains this characteristic and is transformed into
$$T = \frac{m}{2}(\dot{\xi}^2 + \dot{\eta}^2).$$
As the velocities are transformed in the same way as the coördinates we have the condition that the linear transformation must leave the quantity $x^2 + y^2$ invariant, i.e.,
$$x^2 + y^2 = \xi^2 + \eta^2.$$
Such transformations are called "orthogonal." They correspond geometrically to *rotations* of the coördinate system around the origin in the xy-plane, because for such a rotation the distance r or $r^2 = x^2 + y^2$ is in fact invariant. Now an equation of the form
$$a_{11}x^2 + 2\,a_{12}xy + a_{22}y^2 = 2\,U = \text{const.}$$
with a definite left-hand side represents an ellipse with the origin at the center. This ellipse has two principal axes a, b. If these are chosen as $\xi\eta$-axes the equation of the ellipse $\dfrac{\xi^2}{a^2} + \dfrac{\eta^2}{b^2} = 1$ becomes
$$2\,U = \kappa_1\xi^2 + \kappa_2\eta^2,$$

THE STRUCTURE OF THE ATOM

where $a^2 = 2\,U/\kappa_1$ and $b^2 = 2\,U/\kappa_2$, and we have the desired expression. The equations of motion are now

$$m\ddot{\xi} + \kappa_1\xi = 0, \qquad m\ddot{\eta} + \kappa_2\eta = 0$$

and, therefore, the frequencies are

$$\nu_1 = \frac{1}{2\pi}\sqrt{\frac{\kappa_1}{m}} = \frac{1}{2\pi a}\sqrt{\frac{2\,U}{m}}$$

$$\nu_2 = \frac{1}{2\pi}\sqrt{\frac{\kappa_2}{m}} = \frac{1}{2\pi b}\sqrt{\frac{2\,U}{m}}.$$

Similar relations hold for any arbitrary number of degrees of freedom.

Formerly, in order to interpret line spectra, attempts were made to construct mechanical systems which should have just the observed lines as proper frequencies, but none of them gave rise to useful results; in other words, none led to oscillating systems built out of known elementary particles (protons and electrons) governed by known laws or reasonable modifications of them and having these frequencies.

In our new theory the relation between the principal axes of the quadratic form and the frequency enters again, except that instead of the observed frequencies the values of the *terms* or *energy levels* occur. These appear as the reciprocal axes of a certain Hermitian form. The frequencies appear later as differences between terms.

To each matrix $a = (a(nm))$ corresponds a bilinear form

$$A(xy) = \sum_{m,n} a(nm)x_n y_m \tag{2}$$

of two systems of variables. If the matrix is Hermitian,

$$\tilde{a} = a^*, \qquad a(mn) = a^*(nm), \tag{3}$$

where the symbol $\sim$ indicates interchange of rows and columns and the symbol $*$ change to conjugate complex quantities, then the form A assumes real values if we place for the variables y_n the values conjugate to x_n:

$$A(xx^*) = \sum_{n,m} a(nm)x_n x_m^* \text{ is real.} \tag{4}$$

Let us recall the rule, easily proved, that $(\widetilde{ab}) = \widetilde{b}\widetilde{a}$. Applying a linear transformation to x_n,

$$x_n = \sum_l v(ln) y_l \tag{5}$$

with the complex matrix $v = (v(ln))$ the bilinear form A is transformed into

$$A(xx^*) = B(yy^*) = \sum_{n,m} b(nm) y_n y_m^*$$

where

$$b(nm) = \sum_{k,l} v(nk) a(kl) v^*(ml)$$

or in matrix notation

$$b = va\widetilde{v}^*. \tag{6}$$

The matrix b is said to be the transform of a. The matrix b is again Hermitian, for

$$\widetilde{b} = v^* \widetilde{a} \widetilde{v} = v^* a^* \widetilde{v} = b^*. \tag{7}$$

The matrix v is said to be orthogonal if the corresponding transformation leaves the Hermitian unit form

$$E(xx^*) = \sum_n x_n x_n^*$$

invariant. According to the result just obtained this is true when, and only when,

$$v\widetilde{v}^* = 1 \quad \text{or} \quad \widetilde{v}^* = v^{-1}. \tag{8}$$

For a finite number of variables the same theorems hold in general for Hermitian forms as for real quadratic forms. Here also there always exists an orthogonal principal-axis transformation by which A becomes a sum of squares,

$$A(xx^*) = \sum_n W_n y_n y_n^*.$$

For matrices this means that there exists a matrix v for which

$$v\tilde{v}* = 1 \text{ and } va\tilde{v}* = vav^{-1} = W \qquad (9)$$

where $W = (W_n \delta_{mn})$ is a diagonal matrix. A similar theorem exists for infinite matrices in all cases so far investigated. It may happen that, in the right-hand side of these equations, n takes, besides a discrete series of values, a continuous series, to each of which correspond integral components in our formulas. The quantities W_n are called "characteristic values," their totality constitutes the "mathematical" spectrum of the form, consisting of a "point" spectrum and an "interval" spectrum. This spectrum is, as pointed out before and as will be shown presently, identical with the "term spectrum" of physics, while the "frequency spectrum" is obtained from the former by difference relations.

The transformation along the principal axes gives at once the solution of the dynamical problem which can be formulated as follows: Let any system of coördinates and momenta $q_k{}^0$, $p_k{}^0$ be given satisfying the commutation relations, for instance those of a system of uncoupled resonators. A transformation $(q_k{}^0 p_k{}^0) \to (q_k p_k)$ must be found leaving the commutation relations (1) invariant and transforming the energy into a diagonal matrix. According to the theorem above an orthogonal matrix S for which

$$S\tilde{S}* = 1 \qquad S*S = 1$$

exists such that by the transformation,

$$\left. \begin{array}{l} p_k = S p_k{}^0 \tilde{S}* = S p_k{}^0 S^{-1} \\ q_k = S q_k{}^0 \tilde{S}* = S q_k{}^0 S^{-1}, \end{array} \right\} \qquad (10)$$

(1) the Hermitian character of $p_k{}^0$, $q_k{}^0$ is conserved for p_k, q_k,
(2) the commutation relations remain invariant,
(3) the energy is transformed into a diagonal matrix

$$H(pq) = SH(p^0 q^0)S^{-1} = W. \qquad (11)$$

It is important to add that the transforming matrix and the series of W-values may have continuous parts. This has been shown by Hilbert and Hellinger for a certain class of infinite matrices belonging to the so-called "bounded forms." The same must be expected *a priori* of our matrices which in general do not satisfy the condition of bounded forms. A continuous series of energy values W or of terms $\dfrac{W}{h}$ is thus obtained. Accordingly there are evidently three kinds of elements in the coördinate matrices:

(1) Those for which both m and n belong to the discrete series of values of W. These correspond to jumps between periodic orbits and give the line spectrum.

(2) Those for which n belongs to the discrete and m to the continuous series of values of W or conversely. These correspond to jumps between periodic and aperiodic orbits and give those known continuous spectra which exist beyond the limits of line series.

(3) Those for which both n and m belong to the continuous series of W-values. These correspond to jumps between two aperiodic orbits and give the continuous spectrum in the proper sense.

The actual mathematical calculation of the continuous spectrum on the basis of this theory is, however, impossible partly on account of the intricacy of the calculations and more particularly because of difficulties of convergence. The integrals are improper or altogether divergent. This is related to the fact that aperiodic motions approach uniform rectilinear motion asymptotically in the limit of infinite distance. This motion has evidently no period and represents the case of greatest singularity. It is not amenable to matrix representation, even if continuous matrices are mustered for the purpose.

LECTURE 20

Substitution of the matrix calculus by the general operational calculus for improved treatment of aperiodic motions — Concluding remarks.

In the case of aperiodic straight-line motion, another procedure must therefore be adopted, which Wiener and I have recently developed. Only an outline of the fundamental ideas can be given here. An Hermitian form can be associated with every matrix, as already shown; likewise a linear transformation of the form used above,

$$x_n = \sum_l v(ln) y_l. \qquad (1)$$

Then the product of two matrices corresponds to the successive application of two such transformations:

$$x_n = \sum_k q(nk) y_k, \qquad y_k = \sum_m p(km) z_m.$$

These together give

$$x_n = \sum_m qp(nm) z_m, \qquad (2)$$

where

$$qp(nm) = \sum_k q(nk) p(km).$$

As seen, the matrix enters here not as a "quantity" or "system of quantities," but as an *operator* which, from an infinite system of quantities y_1, $y_2 \cdots$, yields another system x_1, $x_2 \cdots$. The precise physical significance of these quantities is still very obscure. A calculus of operators can, therefore, be substituted for the matrix calculus and this method becomes fruitful if applied in the following way: An infinite system of quantities x_1, $x_2 \cdots$ may define a function with a continuous range of arguments; for instance these quantities may be taken as the coefficients of a Fourier series. It is advantageous to operate with this function instead of with the coefficients, because we then have the whole machinery of the calculus at our disposal, and differential or integral equations replace infinite sets of simultaneous equations in an infinite number of variables. These equations possess solutions under certain conditions even when the original representation in series collapses. Of course Fourier series will not be used here, but general trigonometric series of the form

$$x'(t) = \sum_n x_n e^{\frac{2\pi i}{h} W_n t}. \tag{3}$$

The coefficients x_n are determined from the function $x(t)$ by taking averages,

$$x_n = \lim_{T \to \infty} \frac{1}{2T} \int_{-T}^{T} x(s) e^{-\frac{2\pi i}{h} W_n s} ds. \tag{4}$$

Instead of the matrix $q = (q(mn))$ we make use of the function of two variables,

$$q(t, s) = \sum_{mn} q_{mn} e^{\frac{2\pi i}{h}(W_m t - W_n s)} \tag{5}$$

and of the derived "average operator,"

$$q = \left(\lim_{T \to \infty} \frac{1}{2T} \int_{-T}^{T} q(t, s) ds \cdots \right). \tag{6}$$

It is then easy to show that operator products, formed by the successive application of operators, correspond to matrix products. An explicit representation of operators is not, however, necessary. It is sufficient to consider linear operators in general, that is such operators for which the simple formula

$$q(x(t) + y(t)) = qx(t) + qy(t)$$

holds. Thus multiplication by a function of t, differentiation and integration with respect to t, for example, are all operators. Of special importance is the differential operator $D = d/dt$.

Under certain conditions a matrix can be associated with an operator. The sequence of energy levels of this matrix is ordered not with respect to indices m, n, but relatively to the

THE STRUCTURE OF THE ATOM 127

energy values themselves. The elements of the matrix which correspond to the operator q are defined by

$$q(V, W) = \lim_{T \to \infty} \frac{1}{2T} \int_{-T}^{T} e^{-\frac{2\pi i}{h} Vt} q e^{\frac{2\pi i}{h} Wt} dt. \qquad (7)$$

In many cases this matrix does not exist although the sum of elements in a row

$$q(t, W) = e^{-\frac{2\pi i}{h} Wt} q e^{\frac{2\pi i}{h} Wt} \qquad (8)$$

does. For the operator D for instance,

$$q(V, W) = \lim_{T \to \infty} \frac{1}{2T} \int_{-T}^{T} \frac{2\pi i}{h} W e^{\frac{2\pi i}{h}(W-V)t} dt = \begin{cases} \frac{2\pi i}{h} W & \text{if } V = W \\ 0 & \text{otherwise,} \end{cases}$$

$q(V, W)$ therefore does not exist as a continuous function. If W has discrete values, $q(V, W)$ is the diagonal matrix $(W_n \delta_{nm})$. The sum of the elements in a row

$$q(t, W) = \frac{2\pi i}{h} W$$

however, always exists. From this example it is seen how the operational method permits a treatment of singular cases where the matrix representation breaks down.

A more exhaustive treatment of the method is as yet unwarranted. Suffice to say that it has been possible to show that, in the case of the harmonic oscillator, the operational calculus gives the same result as the matrix calculus. Moreover, treatment of uniform rectilinear motion is possible, a case where the matrix calculus breaks down completely. Investigations on the theorems of angular momentum, on hyperbolic orbits in the hydrogen atom and on similar problems are in progress.

In closing I should like to add a few general remarks. The first concerns the question of whether it is possible to visualize the laws of physics as formulated in this new manner and whether the processes in the atom can be conceived to exist in space and time. A definite answer will only be possible when we can see all the consequences of the new theory, perhaps only when new principles have been discovered. But it already seems certain that the usual conceptions of space and time are not rigorously compatible with the character of the new laws.

Consider for instance the hydrogen atom. The classical theory not only gives the orbits of the electron but seems to assign a meaning to the position of the electron at each instant. In the new theory the energy and moment of momentum of a state can be given, but it appears to be impossible to give any further description of this state as a geometrical orbit and even more impossible to fix the position of the electron at any instant. Space points and time points in the ordinary sense do not exist. These conceptions can only be introduced subsequently in limiting cases.

On the other hand it seems to me that we have a right to use the terms "orbit" or even "ellipse," "hyperbola," etc. in the new theory, if we agree to interpret them rationally and to understand by them the quantum processes which go over in the limit to the orbits, ellipses, hyperbolas, etc. of the classical theory. This not only gives a convenient terminology but expresses the following fact: The world of our imagination is narrower and more special in its logical structure than the world of physical things. Our imagination is restricted to a limiting case of possible physical processes. This philosophical point of view is not new: it has always been the guiding thought of physicists since Copernicus, and it came so clearly to the fore in the theory of relativity that philosophy was compelled to take a definite stand towards it. In the quantum theory this guiding principle assumes an even more predominant rôle, but in this case it is supported by such an enormous weight of evidence that a flat denial seems much more difficult than it was when the theory of relativity came up for consideration.

Only a further extension of the theory, which in all likelihood will be very laborious, will show whether the principles given above are really sufficient to explain atomic structure. Even if we are inclined to put faith in this possibility, it must be remembered that this is only the first step toward the solution of the riddles of the quantum theory. Our theory gives the *possible* states of the system, but no indication of whether a

THE STRUCTURE OF THE ATOM

system is in a given state. It gives at most the probability of the jumps. However, the statement that a system is at a given time and place in a certain state probably has a meaning, which the present state of our theory does not allow us to formulate. This is also the case with regard to the problem of light-quanta. Here the Compton effect and the related experiments of Bothe and Geiger, Compton and Simon have shown that both the energy and the momentum of light travel as a projectile from atom to atom. But the existence of interference, that is the fact that light added to light can produce darkness, is just as certain. It cannot yet be seen how these two views can be reconciled or whether a matrix representation of the electromagnetic field will lead further. An attempt to treat the statistics of cavity radiation by the new method has resulted in the elimination of serious contradictions in the classical theory. Many puzzling questions remain which fall outside the scope of these lectures.

In the further development of the new quantum theory, the physicist cannot dispense with the aid of the mathematician. The close alliance between mathematics and physics which has reigned during the best periods of both sciences will, I hope, return and banish the mystic cloud in which physics has of late been enshrouded. The activity of the mathematician must however not carry him as far as in the theory of relativity, where the clarity of his reasoning has come to be hidden by the erection of a structure of pure speculation so vast that it is impossible to view it in its entirety. A single crystal can be clear, nevertheless a mass of fragments of this crystal is opaque. Even the theoretical physicist must be guided by the ideal of the closest possible contact with the world of facts. Only then do the formulas live and beget new life.

SERIES II
THE LATTICE THEORY OF RIGID BODIES

SERIES II
THE LATTICE THEORY OF RIGID BODIES

LECTURE 1

Classification of crystal properties — Continuum and lattice theories — Geometry of lattices.

The theory to be developed in these lectures is an application of the atomic theory of matter, the general foundations of which have been given in the first series of lectures (The Structure of the Atom). This application of the general ideas can be developed independently; it is of a special character and of interest only to those who have to deal with the properties of solid bodies, especially those of crystals. Therefore, it is justifiable to develop this part of the theory by itself. At the beginning of the first series of lectures, a general introduction on the importance of atomic theory for a knowledge of the laws of nature was given from a philosophical standpoint. Here, we shall start with the theory itself.

We have satisfactory theories of gases and solids, but not of fluids. The reason for this can be seen if we consider the packing of atoms from two points of view, *viz.* density and regularity. With our mathematical methods we command two limiting cases: that of extremely small density with any irregularity, as in the case of the ideal gases; and the case of absolute regularity with any density: i.e., the case of the ideal solids, or crystals. We are not able to formulate a satisfactory theory in the case of atoms which are both irregularly and densely packed, as in the case of fluids. Not only have we insufficient knowledge of atomic structure, but the mathematical problems encountered here are too difficult.

In the following series of lectures we shall deal with solids, and start with the ideal, absolutely regular solid. This is the crystal at the zero point of the absolute temperature scale, where we have no disturbances in the regular packing due to heat motion. From this point we can progress to more complicated cases; first, by a consideration of the heat motion, then, of other disturbances. But we shall not go very far in this direction and shall omit the theory of crystalline mixtures and of amorphous solids, which can also be considered as very viscous fluids. We assume that the structure of atoms and the forces acting between the atoms are known. The problem is to calculate the structure and the properties of the crystal made up of these atoms. Actually, atomic structure is not well enough known to solve this problem, and therefore we shall use the relations existing between atomic structure and crystal structure to find out something about atoms from the observed properties of crystals.

All changes that can be made in a body can be reduced to homogeneous changes, that is, changes such that every point of the body is affected in the same way, for any function of the coördinates can be regarded as linear in a sufficiently small region of space.

We shall classify the properties of a crystal from the macroscopic standpoint of a continuous medium and study three properties: (1) Mechanical properties, (2) Electrical properties, (3) Thermal properties. The magnetic properties, which are not so important, we shall leave aside. Each of these properties is determined by two sorts of quantities: extensive and intensive. The following figure, after Heckmann, will illustrate what is meant (Fig. 17).

We have two concentric triangles. In the corners of the inner one we have the extensive quantities: strain, electric polarization, and temperature; in the corners of the outer triangle the intensive quantities: stress, electric field, and heat energy. Between these points are written the phenomena connecting them. The primary phenomena are elasticity, the interaction between strain and stress; dielectricity, the interaction between electric field and polarization; and specific heat, the interaction between temperature and heat content. The secondary phenomena with the inverse relation are: thermal expansion and heat of deformation, piezoelectricity and electrostriction, pyroelectricity and electrocaloric effect. This is the material to be explained by the mathematical theory of crystals.

There are two steps in the explanation. The formal connections given in our figure can be deduced without exact knowledge of the laws of interaction of the atoms, as we are concerned with small displacements only. Therefore, we can

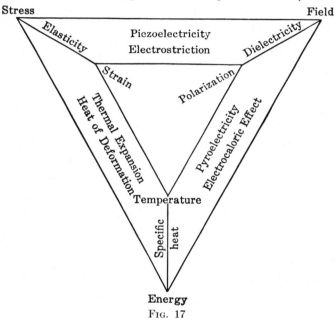

FIG. 17

develop the functions representing atomic forces in series in powers of small displacements. Then we get the physical phenomena expressed by the coefficients of these developments; they enter in the formulas as unknown parameters and have to be determined by experiment. This part of the investigation we call the formal theory of crystals. I shall give you a short sketch of it, chiefly to show where the atomic theory leads farther than the old continuum theory.

The second and more important step will be to make use of

the properties of atoms and atomic forces, which are more or less well known from other parts of physics, to calculate these coefficients (for instance the constants of elasticity, of piezoelectricity, etc.). This problem has been attacked hitherto only for a small class of crystals, namely, those built up of ions. In this case, the electrical forces originating from the charge of the ions and following the well-known Coulomb law outweigh all other less well-known forces so much that the latter can be regarded as small corrections. The results of this theory will be presented later.

Let us now begin with the mathematical description of a crystal lattice. Each lattice can be thought of as built up of a group of atoms, which we call the "basis," by translation in three directions of space. These translations are given by the vectors

$$\mathbf{a}_1, \mathbf{a}_2, \mathbf{a}_3.$$

They determine the elementary parallelepipedon, or the "cell" of the lattice. The volume of the cell is

$$\begin{vmatrix} a_{1x} & a_{1y} & a_{1z} \\ a_{2x} & a_{2y} & a_{2z} \\ a_{3x} & a_{3y} & a_{3z} \end{vmatrix} = \delta^3 = \Delta,$$

where a_{1x} is the scalar component of $\mathbf{a}_1$ along the x-axis and δ is a measure of the linear dimension of the lattice; we shall call δ the "lattice constant."

If $\mathbf{r}$ is the vector from any origin O to any point in the cell, this point moves by the translation of the lattice to an infinite number of equivalent points given by the vectors

$$\mathbf{r} + \mathbf{r}^l, \quad \text{where} \quad \mathbf{r}^l = l_1 \mathbf{a}_1 + l_2 \mathbf{a}_2 + l_3 \mathbf{a}_3.$$

Here l_1, l_2, l_3, are integral numbers (between $-\infty$ and $+\infty$, including zero) which we shall denote by one index l written above, called the "cell-index."

The positions of the atoms of the basis are given by a series of vectors

$$\mathbf{r}_1, \mathbf{r}_2, \cdots \mathbf{r}_k, \cdots \mathbf{r}_s$$

$k = 1, 2, \cdots s$ we call the "basis-index."

THE LATTICE THEORY OF RIGID BODIES

Then any point of the lattice can be denoted by the vector

$$\mathbf{r}_k^l = \mathbf{r}_k + \mathbf{r}^l,$$

the rectangular coördinates of which are

$$\begin{cases} x_k^l = x_k + l_1 a_{1x} + l_2 a_{2x} + l_3 a_{3x} \\ \cdot = \ldots\ldots\ldots\ldots \\ \cdot = \ldots\ldots\ldots\ldots \end{cases}$$

The vector from any point of the basis k' to any other point of the lattice k, l, is given by the symbol:

$$\mathbf{r}_{kk'}^l = \mathbf{r}_k^l - \mathbf{r}_{k'} = \mathbf{r}_k - \mathbf{r}_{k'} + \mathbf{r}^l,$$

and the vector from any point of the lattice to any other is written

$$\mathbf{r}_k^l - \mathbf{r}_{k'}^{l'} = \mathbf{r}_{kk'}^{l-l'}.$$

The masses of the atoms, some of which may be equal, we denote by $m_1, m_2, \cdots m_s$. The whole mass of the basis is then

$$m_1 + m_2 + \cdots + m_s = m$$

and the density

$$\rho = \frac{m}{\Delta}.$$

We shall use three kinds of summations:

(1) sums over the basis index $k = 1, 2, \ldots\ldots$ We denote these by

$$\underset{k}{\Sigma}, \underset{kk'}{\Sigma}, \cdots$$

(2) sums over terms originating one from the other by cyclic changes of coördinates, which we denote by

$$\underset{x}{\Sigma}, \underset{xy}{\Sigma}, \cdots,$$

for instance, the scalar product of two vectors

$$\mathbf{a} \cdot \mathbf{b} = a_x b_x + a_y b_y + a_z b_z = \underset{x}{\Sigma} a_x b_x$$

(3) sums over the cell-index l. We write these as

$$\underset{l}{S}, \underset{l'}{S}, \cdots$$

where the indices $l_1, l_2, l_3, \cdots$ can in general assume values from $-\infty$ to $+\infty$.

Limits of summation we write as inequalities, i.e.,

$$\underset{l \geq 0}{S}$$

LECTURE 2

Molecular forces — Polarizability of atoms — Potential energy and inner forces — Homogeneous displacements — The conditions of equilibrium — Examples of regular lattices.

In general we shall suppose that the forces between two atoms of the lattice are central, but in many cases this hypothesis is not sufficient. For example, we consider three atoms the electronic structure of which is so weak that they are noticeably polarized by the influence of their fields of force. By this process potential energy is stored up in the atom. We may equate the resulting electric moment **p** to $e\mathbf{u}$ where **u** is the mean displacement of the charge e. Supposing that the charge is quasi-elastically attached to its position of equilibrium, this energy becomes proportional to u^2. Therefore it is also proportional to p^2. We equate it to $\dfrac{p^2}{2\,\alpha}$. If now **E** is the external field which produces the polarization, it does the work

$$e\mathbf{E}\cdot\mathbf{u} = \mathbf{p}\cdot\mathbf{E}$$

on the charge e during the translation **u** and therefore, the total energy of the polarized atom in the field **E** is

$$\mathbf{p}\cdot\mathbf{E} + \frac{p^2}{2\,\alpha}. \qquad (1)$$

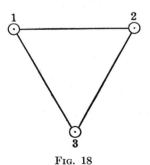

Fig. 18

In our case of three atoms, the field **E** is produced by the charges of the other atoms. For instance, for the atom 3

$$\mathbf{E}_3 = \frac{e_1\mathbf{r}_{13}}{r_{13}{}^3} + \frac{e_2\mathbf{r}_{23}}{r_{23}{}^3}.$$

Denoting by $\phi_{12}(r)\cdots$ those parts of the potential energy which come from the radial forces, the total potential energy of the system is

$$\Phi = \phi_{23}(r_{23}) + \phi_{31}(r_{31}) + \phi_{12}(r_{12})$$
$$+ \mathbf{p}_1 \cdot \left(\frac{e_2 \mathbf{r}_{21}}{r_{21}^3} + \frac{e_3 \mathbf{r}_{31}}{r_{31}^3}\right) + \mathbf{p}_2 \cdot \left(\frac{e_3 \mathbf{r}_{32}}{r_{32}^3} + \frac{e_1 \mathbf{r}_{12}}{r_{12}^3}\right) + \mathbf{p}_3 \cdot \left(\frac{e_1 \mathbf{r}_{13}}{r_{13}^3} + \frac{e_2 \mathbf{r}_{23}}{r_{23}^3}\right)$$
$$+ \frac{p_1^2}{2\,\alpha_1} + \frac{p_2^2}{2\,\alpha_2} + \frac{p_3^2}{2\,\alpha_3}. \tag{2}$$

The conditions of equilibrium demand that the derivatives of Φ with respect to the coördinates and along the components of $\mathbf{p}$ vanish. These conditions may be used to eliminate $\mathbf{p}$ from the expression of Φ. We have, for instance

$$\frac{\partial \Phi}{\partial p_{1x}} = \frac{e_2 x_{21}}{r_{21}^3} + \frac{e_3 x_{31}}{r_{31}^3} + \frac{p_{1x}}{\alpha_1} = 0, \cdots$$

whence it follows that

$$\mathbf{p}_1 = -\alpha_1 \left(\frac{e_2 \mathbf{r}_{21}}{r_{21}^3} + \frac{e_3 \mathbf{r}_{31}}{r_{31}^3}\right), \cdots \tag{3}$$

and if we introduce this value in Φ

$$\Phi = \phi_{23}(r_{23}) + \phi_{31}(r_{31}) + \phi_{12}(r_{12})$$
$$- \frac{\alpha_1}{2}\left(\frac{e_2 \mathbf{r}_{21}}{r_{21}^3} + \frac{e_3 \mathbf{r}_{31}}{r_{31}^3}\right)^2 - \frac{\alpha_2}{2}\left(\frac{e_3 \mathbf{r}_{32}}{r_{32}^3} + \frac{e_1 \mathbf{r}_{12}}{r_{12}^3}\right)^2 - \frac{\alpha_3}{2}\left(\frac{e_1 \mathbf{r}_{13}}{r_{13}^3} + \frac{e_2 \mathbf{r}_{23}}{r_{23}^3}\right)^2. \tag{4}$$

This expression cannot be decomposed additively into terms which depend only on the three distances r_{12}, r_{23}, r_{31}, respectively, for it depends also on the scalar products $\mathbf{r}_{21} \cdot \mathbf{r}_{31}, \cdots$, that is, on the angles of the triangle. Therefore we are no longer concerned with simple central forces, but it is very easy to transform Φ so that only distances, and no angles, appear. We have $\mathbf{r}_{23} + \mathbf{r}_{31} + \mathbf{r}_{12} = 0$ whence there follows, for instance, the relation (theorem of Pythagoras)

$$r_{23}^2 = r_{31}^2 + r_{12}^2 + 2\,\mathbf{r}_{31} \cdot \mathbf{r}_{12}$$

by means of which the scalar products can be expressed in terms of the distances. We can therefore suppose that Φ has the form

$$\Phi(r_{23}, r_{31}, r_{12})$$

but not
$$\Phi = \phi_{23}(r_{23}) + \phi_{31}(r_{31}) + \phi_{12}(r_{12}).$$

An analogous relation holds in the case of more than three atoms. I might add here that the general law of force is necessary in many cases, and it has recently been applied by several of my associates. For the sake of simplicity we restrict ourselves to radial forces.

The number of atoms in a crystal lattice is extremely large, as the distance between atoms is of the order of 10^{-8} cm. Therefore we shall always consider crystals as of infinite extent. This sometimes raises difficulties as to the convergence of the series required in the calculation of the potential energy and it is necessary to compute the summations by special methods.

Let the potential energy of any two atoms of the kinds k and k' be $\phi_{kk'}(r)$. If the atom k' lies in the base cell, and the atom k in the cell l, then the distance between them at equilibrium is $|\mathbf{r}^l_{kk'}|$. Let us denote the corresponding energy by $\phi^l_{kk'} = \phi_{kk'}(|\mathbf{r}^l_{kk'}|)$. Now were we to express this total energy of the lattice as a double sum over the infinite lattice we would not obtain a finite value. Therefore we proceed to find the energy of all points of the lattice with respect to one point k' of the base

$$S\underset{l\ k}{\Sigma}\phi^l_{kk'}$$

and sum over the base. The resulting expression

$$\phi_0 = S\underset{l\ kk'}{\Sigma}\phi^l_{kk'}$$

is evidently independent of which cell we have chosen as base. Further, the sum will be convergent if the functions $\phi_{kk'}(r)$ decrease rapidly enough with respect to r. ϕ_0 denotes the energy of all atoms of the lattice with respect to all those of any cell. Accordingly the energy of the lattice which is composed of a great number N of cells is very nearly

$$\Phi_0 = \tfrac{1}{2} N \phi_0$$

where the factor $\tfrac{1}{2}$ is introduced because without it the energy of every pair of cells would be counted twice. The expression is only approximate in so far as the influence of the surface of the crystal, which is in fact finite, is not considered in the summation over ϕ_0. This must be taken into account by special considerations. As in the continuous-medium theory, surface energy is considered in a separate chapter on capillarity. The magnitude of Φ_0 is a function of the components of the vectors $\mathbf{a}_1, \mathbf{a}_2, \mathbf{a}_3, \mathbf{r}_1, \mathbf{r}_2, \cdots \mathbf{r}_s$

$$\Phi_0 = \Phi_0(\mathbf{a}_1 \cdots \mathbf{r}_1 \cdots \mathbf{r}_s).$$

It is of course an orthogonal invariant, that is, it remains unchanged not only by translations, but also by rotations of the lattice.

Now we consider disturbances of the equilibrium. To every point of the lattice we give a displacement $\mathbf{u}_k^l$ and denote the relative displacements of two lattice points by

$$\mathbf{u}_{kk'}^{ll'} = \mathbf{u}_k^l - \mathbf{u}_{k'}^{l'};$$

then the vector from one lattice point to another changes from

$$\mathbf{r}_{kk'}^{l \cdot l'} \text{ to } \mathbf{\bar{r}}_{kk'}^{l \cdot l'} = \mathbf{r}_{kk'}^{l \cdot l'} + \mathbf{u}_{kk'}^{ll'}$$

and its components are to be substituted for the coördinates in the expression for Φ_0.

We shall first consider an especially simple class of displacements, which we may call "homogeneous disturbances." These consist of small changes made in the elements of the cell and of the base, the lattice being rebuilt out of these altered elements. a_{1x} must therefore be replaced by

$$\bar{a}_{1x} = a_{1x} + \Sigma u_{xy} a_{1y}, \cdots$$

and x_k by

$$\bar{x}_k = x_k + u_{kx} + \Sigma_y u_{xy} y_k, \cdots.$$

Here the displacement is separated into two parts, one a homogeneous disturbance of the form and the contents of the cell, given by the tensor u_{xy}, the other a displacement of the simple lattices k as rigid structures given by the vectors $\mathbf{u}_k$. If u_{xy} is arbitrary the first part also comprises a rotation of the whole lattice. To exclude this we must have $u_{xy} = u_{yx}$.

THE LATTICE THEORY OF RIGID BODIES 143

Writing the expression for the potential energy $\bar{\Phi}_0$ of the lattice, as constructed from the altered base and cell, in terms of $\bar{a}_1, \bar{a}_2, \bar{a}_3, \bar{r}_1 \cdots \bar{r}_s$, $\bar{\Phi}_0$ becomes dependent not only on the equilibrium coördinates of the undisturbed lattice, but also on the components of the displacements u_{xy}, u_{kx}. For equilibrium it is necessary that the terms of the first order in the expansion of $\bar{\Phi}_0$ in series with respect to the displacements $u_{xy}, u_{kx} \cdots$,

$$\bar{\Phi}_0 = \Phi_0 + \sum_{kx}\left(\frac{\partial \bar{\Phi}_0}{\partial u_{kx}}\right)_0 u_{kx} + \sum_{xy}\left(\frac{\partial \bar{\Phi}_0}{\partial u_{xy}}\right)_0 u_{xy} + \cdots, \qquad (5)$$

vanish. This gives us two kinds of equations

$$K^0_{kx} = -\left(\frac{\partial \bar{\Phi}_0}{\partial u_{kx}}\right)_0 = 0$$

$$K^0_{xy} = -\left(\frac{\partial \bar{\Phi}_0}{\partial u_{xy}}\right)_0 = 0. \qquad (6)$$

Not all of these equations are independent, because if the whole lattice is moved as a rigid body (that is, if all the $\mathbf{u}_k$ are equal, and all the $u_{xy} = 0$), or if it is rotated (that is, if $u_{xx} = u_{yy} = u_{zz} = 0$ and $u_{yz} + u_{zy} = 0 \cdots$ and all the $\mathbf{u}_k = 0$) the potential energy must identically vanish. Hence the following identities must hold:

$$\begin{cases} \sum_k K^0_{kx} = 0 \cdots \\ K^0_{yz} = K^0_{zy} \cdots . \end{cases}$$

Evidently the vectors $\mathbf{K}^0_k$ represent the forces which the simple elements of the lattice exert on each other in a homogeneous disturbance and the quantities K^0_{xy} represent the components of the tension tensor. Both vanish at equilibrium.

The number of these equations is just sufficient to calculate the elements $\mathbf{a}_1, \mathbf{a}_2, \mathbf{a}_3, \mathbf{r}_1, \cdots \mathbf{r}_s$ of the lattice in so far as they have a physical significance, that is, apart from any translations or rotations. If the number of the elements is diminished by properties of symmetry, the number of equations is diminished in the same way. In regular crystals, for instance, we can choose cubic cells. Here everything is determined by symmetry except the length of the side $\delta = |\mathbf{a}_1| = |\mathbf{a}_2| = |\mathbf{a}_3|$. Therefore the number of stress equations is reduced to one. This can be written in the form

$$\frac{d\Phi_0}{d\delta} = 0. \tag{7}$$

As an example we shall consider here, and later, the elementary law of the form (Fig. 19)

$$\phi_{kk'}(r) = -\frac{a_{kk'}}{r^m} + \frac{b_{kk'}}{r^n} \qquad (a_{kk'},\ b_{kk'} > 0,\ m < n)$$

which consists of one (negative) term of attraction, and one (positive) term of repulsion.

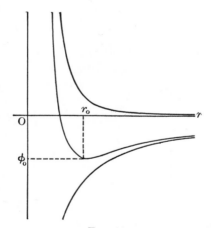

Fig. 19

By summation we obtain

$$\Phi_0 = -\frac{A}{\delta^m} + \frac{B}{\delta^n}. \tag{8}$$

In lattices of ions, for instance, $m = 1$ (Coulomb's potential). $n = \infty$ would mean that the atoms behave as rigid spheres.

THE LATTICE THEORY OF RIGID BODIES 145

The condition for equilibrium becomes

$$\frac{d\Phi_0}{d\delta} = +\frac{mA}{\delta^{m+1}} - \frac{nB}{\delta^{n+1}} = 0$$

and can be used for the elimination of the constant B,

$$\frac{B}{\delta_0^n} = \frac{m}{n}\frac{A}{\delta_0^m},$$

therefore

$$\Phi_0 = -\frac{n-m}{n}\frac{A}{\delta_0^m}. \tag{9}$$

LECTURE 3

Elimination of inner motions — Compressibility — Elasticity and Hooke's law — Cauchy's relations — Dielectric displacement and piezoelectricity — Residual-ray frequencies.

We shall now consider the terms of the second order in the potential energy. They have the form

$$\overline{\Phi}_0 = \Phi_0 + \tfrac{1}{2}\left\{\sum_{kl}\sum_{xy}\left(\frac{\partial^2 \overline{\Phi}_0}{\partial u_{kx}\partial u_{ly}}\right)_0 u_{kx}u_{ly}\right.$$
$$+ 2\sum_{k}\sum_{xyz}\left(\frac{\partial^2 \overline{\Phi}_0}{\partial u_{kx}\partial u_{yz}}\right)_0 u_{kx}u_{yz}$$
$$\left.+ \sum_{xy\,\bar{x}\bar{y}}\left(\frac{\partial^2 \overline{\Phi}_0}{\partial u_{xy}\partial u_{\bar{x}\bar{y}}}\right)_0 u_{xy}u_{\bar{x}\bar{y}}\right\} + \cdots \quad (1)$$

In this formula are included all effects which are due to mechanical and electrical disturbances of equilibrium at the zero point of temperature. That is, those included in the upper part of our diagram:

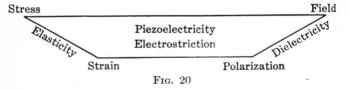

Fig. 20

Let us first consider regular crystals. We shall consider them subjected to a uniform dilatation, by which δ only is altered. Then the equation above reduces to

$$\overline{\Phi}_0 = \Phi_0 + \tfrac{1}{2}\left(\frac{d^2\overline{\Phi}_0}{d\delta^2}\right)_0 (\delta - \delta_0)^2 + \cdots \quad (2)$$

Now the compressibility is defined by

$$\kappa = -\frac{1}{\Delta_0}\frac{\Delta - \Delta_0}{p - p_0} = -\frac{1}{\delta_0^3}\frac{\delta^3 - \delta_0^3}{p - p_0} = -\frac{3}{\delta_0}\frac{\delta - \delta_0}{p - p_0} \quad (3)$$

THE LATTICE THEORY OF RIGID BODIES 147

where p denotes the pressure. But $(p - p_0)$ is given by

$$p - p_0 = -\frac{d\overline{\Phi}_0}{d\Delta} = -\frac{1}{3\,\delta_0{}^2}\frac{d\overline{\Phi}_0}{d\delta} = -\frac{1}{3\,\delta_0{}^2}\left(\frac{d^2\overline{\Phi}_0}{d\delta^2}\right)_0 (\delta - \delta_0)$$

whence we obtain

$$\kappa = \frac{9\,\delta_0}{\left(\dfrac{d^2\overline{\Phi}_0}{d\delta^2}\right)_0}. \tag{4}$$

From the law of force given above, (Equation 8, Lecture 2)

$$\Phi_0 = -\frac{A}{\delta^m} + \frac{B}{\delta^n},$$

we have

$$\frac{d^2\Phi_0}{d\delta^2} = -\frac{m(m+1)A}{\delta^{m+2}} + \frac{n(n+1)B}{\delta^{n+2}}.$$

Eliminating B as above we obtain

$$\frac{d^2\Phi_0}{d\delta^2} = m(n-m)\frac{A}{\delta_0{}^{m+2}}$$

whence follows the relation between the energy of the lattice and the compressibility, first given by Grüneisen:

$$\Phi_0 = -\frac{9\,\delta_0{}^3}{nm}\frac{1}{\kappa}. \tag{5}$$

For a monatomic lattice consisting of neutral atoms, $-\Phi_0$ is the energy of sublimation per atom, and therefore directly measurable. This equation can be used to estimate the value of the product $m \times n$. For the determination of the separate factors m and n we need further data. For ions, where m is known ($=1$), Φ_0 does not have a simple physical meaning but it is possible to connect it with directly measurable quantities as we shall show later.

We shall now study the conditions in the case of general disturbances of regular diatomic crystals of an especially simple type. I have called this type a "diagonal lattice," because all the atoms of the base can be chosen on the diagonal of the cubic cell. To this type belong the well known lattices of NaCl (rock salt), CsCl (cæsium chloride), ZnS (zinc blende) and others. The potential energy per unit volume has here the form

$$U = \frac{\Phi_0}{N\Delta} = \frac{\phi_0}{2\Delta} = \frac{A}{2}(x_x^2 + y_y^2 + z_z^2) + B(y_y z_z + z_z x_x + x_x y_y)$$
$$+ \frac{B}{2}(y_z^2 + z_x^2 + x_y^2) - C\{y_z(u_{1x} - u_{2x}) + z_x(u_{1y} - u_{2y})$$
$$+ x_y(u_{1z} - u_{2z})\} + \frac{D}{2}(\mathbf{u}_1 - \mathbf{u}_2)^2, \tag{6}$$

where $x_x = u_{xx}, \cdots$ and $y_z = z_y = u_{yz} + u_{zy}, \cdots$ are the components of the deformation, $\mathbf{u}_1, \mathbf{u}_2$ are the displacement vectors of the two kinds of atoms, and A, B, C, D are constants which may be expressed as summations over the lattice. The elastic stresses are given by

$$-X_x = \frac{\partial U}{\partial x_x}, \cdots$$
$$-Y_z = \frac{\partial U}{\partial y_z}, \cdots.$$

If the terms in $\mathbf{u}_1, \mathbf{u}_2$ are neglected in the expression of U, Hooke's law is obtained in the form

$$-X_x = Ax_x + B(y_y + z_z), \cdots$$
$$-Y_z = By_z, \cdots.$$

We would have therefore only *two* elastic constants. This is the result found by Cauchy in 1828 on the basis of his atomic theory of rigid bodies. The continuous-medium theory of elasticity, however, gives, as is known, three constants for regular crystals. In Voigt's notation:

$$-X_x = c_{11}x_x + c_{12}(y_y + z_z), \cdots$$
$$-Y_z = c_{44}y_z, \cdots.$$

We obtain therefore one of Cauchy's relations, namely

$$c_{12} = c_{44}.$$

THE LATTICE THEORY OF RIGID BODIES

In the general case of a trigonal crystal where Hooke's law,

$$\begin{cases} -X_x = c_{11}x_x + c_{12}y_y + c_{13}z_z + c_{14}y_z + c_{15}z_x + c_{16}x_y \\ \cdots\cdots\cdots\cdots\cdots\cdots\cdots\cdots\cdots\cdots\cdots\cdots \\ \cdots\cdots\cdots\cdots\cdots\cdots\cdots\cdots\cdots\cdots\cdots\cdots \\ -Y_z = c_{41}x_x + c_{42}y_y + c_{43}z_z + c_{44}y_z + c_{45}z_x + c_{46}x_y \\ \cdots\cdots\cdots\cdots\cdots\cdots\cdots\cdots\cdots\cdots\cdots\cdots \\ \cdots\cdots\cdots\cdots\cdots\cdots\cdots\cdots\cdots\cdots\cdots\cdots \end{cases}$$

$$c_{ik} = c_{ki},$$

has 21 constants, the number of Cauchy relations is 6.

Cauchy's atomic theory is not satisfactory inasmuch as it gives more special results than the continuous-medium theory, and later measurements made principally by Voigt have shown that the latter corresponds to the facts. Therefore, Poisson and Voigt generalized the original theory of Cauchy which operated with point centers of force and conceived the atoms as rigid bodies free to turn about their centers. By a complicated method they succeeded in avoiding Cauchy's relations. But it is not necessary to give up the point centers of force if we only consider the relative displacements of the simple lattices one with respect to the other. Then the stress equations are

$$-X_x = Ax_x + B(y_y + z_z), \cdots$$
$$-Y_z = By_z - C(u_{1x} - u_{2x}), \cdots \qquad (7)$$

to which the force equations

$$-K_{1x} = \frac{\partial U}{\partial u_{1x}} = -Cy_z + D(u_{1x} - u_{2x}) \cdots$$

$$-K_{2x} = \frac{\partial U}{\partial u_{2x}} = Cy_z - D(u_{1x} - u_{2x}) \cdots$$

must be added.

For pure elastic disturbances these forces vanish, and we have

$$u_{1x} - u_{2x} = \frac{C}{D}y_z.$$

If we introduce this in the components of the stress-tensor we obtain

$$-X_x = Ax_x + B(y_y + z_z), \cdots$$
$$-Y_z = \left(B - \frac{C^2}{D}\right)y_z, \cdots \qquad (8)$$

or

$$c_{11} = A, \quad c_{12} = B, \quad c_{44} = \left(B - \frac{C^2}{D}\right)$$

and we therefore do not obtain Cauchy's relation

$$c_{12} = c_{44}.$$

In *diatomic* diagonal lattices with central symmetry (types NaCl and CsCl) the constant $C = 0$, as easily seen. Therefore Cauchy's relation holds here, as is shown experimentally for rock salt and sylvite. For the type ZnS, however, we have three constants, in agreement with experiment. For the monatomic metals Cauchy's relations have not been verified. This shows that the structure of metals cannot be considered as a simple lattice.

We can easily develop the other electromechanical effects. Piezoelectricity results from the internal disturbances of the ion lattices one with respect to the other. In our example the disturbance produces an electric moment per unit of volume

$$p_x = \frac{Ze}{\Delta}(u_{1x} - u_{2x}) = \frac{Ze}{\Delta}\frac{C}{D}y_z = e_{14}y_z.$$

Ze is the charge of the ion and $e_{14} = \dfrac{Ze}{\Delta}\dfrac{C}{D}$ is the piezoelectric constant in Voigt's notation. In general, the relation between electric moment and deformation is represented by

$$p_x = e_{11}x_x + e_{12}y_y + e_{13}z_z + e_{14}y_z + e_{15}z_x + e_{16}x_y,$$
$$\cdots\cdots\cdots\cdots\cdots\cdots\cdots\cdots\cdots\cdots\cdots\cdots$$
$$\cdots\cdots\cdots\cdots\cdots\cdots\cdots\cdots\cdots\cdots\cdots\cdots$$

A lattice in which Cauchy's relations hold cannot of course be piezoelectric, but the converse does not hold. A lattice with

THE LATTICE THEORY OF RIGID BODIES 151

inner disturbances need not be piezoelectric. This is exemplified by fluorspar CaF_2, the octahedral planes of which are represented in Fig. 21. The symmetry of the arrangement acts so that two equivalent planes always have equal and opposite displacements. Therefore no electric moment can appear. Besides this vectorial piezoelectricity there exists a tensorial piezoelectricity which depends upon quadrupoles. Voigt first predicted theoretically the existence of this effect and his conclusions were made probable by experiment. In our theory we can easily calculate this effect, but we shall not do so, since quantititave measurements are as yet lacking.

Fig. 21

Finally let us consider dielectric phenomena. If we place the crystal in an electric field the forces $\mathbf{K}_1$, $\mathbf{K}_2$ (per unit of volume) no longer vanish, but are given by

$$\mathbf{K}_1 = -\frac{Ze}{\Delta}\mathbf{E}, \qquad \mathbf{K}_2 = +\frac{Ze}{\Delta}\mathbf{E}.$$

We obtain for the undeformed state the condition

$$\mathbf{u}_1 - \mathbf{u}_2 = \frac{Ze}{\Delta}\frac{1}{D}\mathbf{E}.$$

The electric moment produced by the field is therefore

$$\mathbf{p} = \frac{e}{\Delta}(\mathbf{u}_1 - \mathbf{u}_2) = \frac{Z^2 e^2}{\Delta^2}\frac{1}{D}\mathbf{E} = \frac{\epsilon - 1}{4\pi}\mathbf{E}$$

and the dielectric constant

$$\epsilon = 1 + \frac{4\pi Z^2 e^2}{\Delta^2}\frac{1}{D}. \tag{9}$$

Closely connected to this are the so-called residual rays, that is, those frequencies proper to the lattice discovered by Rubens in the extreme infra-red. To find them we must introduce in the computations the inertia forces:

$$\mathbf{K}_1 = m_1\ddot{\mathbf{u}}_1 - \frac{Ze}{\Delta}\mathbf{E}$$

$$\mathbf{K}_2 = m_2\ddot{\mathbf{u}}_2 + \frac{Ze}{\Delta}\mathbf{E}$$

where m_1, m_2 are the masses of the two kinds of atoms. If now an electric wave the wave-length of which is large compared with the lattice constant, as is the case in the infra-red, acts on the crystal, we can consider the field $\mathbf{E}$ constant in space but periodic in time

$$\mathbf{E} \propto e^{i\omega t}.$$

The equations of motion

$$m_1\ddot{\mathbf{u}}_1 - \frac{Ze}{\Delta}\mathbf{E} + D(\mathbf{u}_1 - \mathbf{u}_2) = 0$$

$$m_2\ddot{\mathbf{u}}_2 + \frac{Ze}{\Delta}\mathbf{E} - D(\mathbf{u}_1 - \mathbf{u}_2) = 0$$

can therefore be solved by setting

$$\mathbf{u}_1, \mathbf{u}_2 \propto e^{i\omega t}.$$

Introducing this in the equations above and subtracting the equations for the x-components one from the other, we obtain

$$\left[D\left(\frac{1}{m_1} + \frac{1}{m_2}\right) - \omega^2\right](u_{1x} - u_{2x}) - \frac{Ze}{\Delta}\left(\frac{1}{m_1} + \frac{1}{m_2}\right)E_x = 0.$$

For $\mathbf{E} = 0$ we have therefore the proper frequency

$$\omega_0 = \sqrt{D\left(\frac{1}{m_1} + \frac{1}{m_2}\right)} \qquad (10)$$

and the action of the light-wave on the lattice is given by:

$$u_{1x} - u_{2x} = \frac{Ze}{\Delta}\left(\frac{1}{m_1} + \frac{1}{m_2}\right)\frac{E_x}{\omega_0^2 - \omega^2}.$$

If we calculate the electric moment $\mathbf{p}$ per unit volume and set

$$\mathbf{p} = \frac{n^2 - 1}{4\pi}\mathbf{E}$$

where n is the index of refraction, we obtain

$$n^2 = 1 + \frac{4\pi Z^2 e^2}{\Delta^2}\left(\frac{1}{m_1} + \frac{1}{m_2}\right)\frac{1}{\omega_0^2 - \omega^2}. \quad (11)$$

For $\omega = 0$ we obtain from this formula the static dielectric constant $n^2 \to \epsilon$. For $\omega \neq 0$ we have the general type of dispersion formula. These formulas were first found by Dehlinger. We have in all 6 physical constants c_{11}, c_{12}, c_{44}, e_{14}, ω_0, ϵ expressed by 4 atomic constants A, B, C, D. We expect therefore that for crystals of the type considered there exist 2 relations among the constants. One of these relations is the equation which we obtain from the dispersion formula by setting $\omega = 0$, $n^2 = \epsilon$. The other is

$$\frac{\epsilon - 1}{4\pi}(c_{12} - c_{44}) = e_{14}^2. \quad (12)$$

If we now ask whether these relations are verified in nature, it is necessary to keep in mind our assumptions. We have considered the atoms as charged mass-points. Therefore we can only expect agreement with experiments if this condition is fulfilled. All processes are excluded which produce noticeable deformations in the atoms themselves, that is, all the processes in which the constant C of internal displacements enters. This constant does not appear in the first identity. However we must keep in mind that the electric field **E** deforms directly the electronic shells of the atoms. If we denote that part of the dielectric constant due to this effect by ϵ_0 and if we replace $(\epsilon - 1)$ in our formula (11) by $(\epsilon - \epsilon_0)$ we obtain

$$\frac{\omega_0^2}{4\pi}(\epsilon - \epsilon_0) = \frac{Z^2 e^2}{\Delta^2}\left(\frac{1}{m_1} + \frac{1}{m_2}\right). \quad (13)$$

This formula is verified in the case of the alkali halides and zinc blende, but not by the salts of silver and thallium. The second relation (12), which contains the piezoelectric constant e_{14}, has been applied to zinc blende, but does not fit the facts as was to be expected on account of the presence of C. To improve the theory Heckmann, at my suggestion, has systematically taken into account the deformability of the ions. The constant of deformability α which we have defined above is well known from the investigations of Heydweiller and Wasastjerna, Fajans and Joos, and Heisenberg and myself. The work of Heckmann has shown that the deformability of the ions has an important effect on the phenomena, and by taking account of it the new formula is more nearly correct. The formulas are so sensitive to small changes in the mechanism that the results of the theory are only qualitative.

LECTURE 4

Ionic lattices — Kossel's and Lewis' theory — Calculation of the lattice energy according to Madelung and Ewald.

The phenomena studied above include in a rough way and in accordance with our diagram, all those which can be calculated from a consideration of homogeneous displacements taking into account terms up to the second order in the expansion of the potential energy. The next step is the study of arbitrary inhomogeneous displacements. The first question here is whether the conditions of equilibrium which we have given are sufficient to annul the linear terms in the potential energy. This is actually the case, as might be expected, because the number of equilibrium conditions is just sufficient to fix the elements of the lattice. Then we have to study the second order terms in the expansion of Φ. For this purpose we first discuss regular wave motion in the lattice and then irregular thermal motion, which, by making use of Fourier series, can be reduced to the former. But before taking up this formal development it will be convenient to give a more concrete meaning to the formulas obtained by applying them to lattices for which we can guess the law of force with fair certainty. At present this is possible only for pure ionic lattices. We now understand the existence of such lattices on the basis of hypotheses concerning the periodic system of the elements developed independently by Kossel in Germany and by Lewis and Langmuir in the United States. We consider the elements in the periodic system which are near the inert gases, for example,

$$\cdots\cdots \text{O} \quad \text{F} \quad Ne \quad \text{Na} \quad \text{Mg} \quad \text{Al} \cdots\cdots$$
$$\cdots\cdots \text{S} \quad \text{Cl} \quad Ar \quad \text{K} \quad \text{Ca} \quad \text{Sc} \cdots\cdots$$

Experiment shows us that the elements F, Cl, etc. easily form monovalent negative ions, the elements Na, K, etc. monovalent

156   PROBLEMS OF ATOMIC DYNAMICS

positive ions. Correspondingly, the elements O, S, etc. and Mg, Ca, etc. form bivalent ions. The theory explains this by the hypothesis that the inert gases have very stable outer electron shells, which, for instance for Ne, Ar, contain 8 electrons. The neighboring elements on the left lack 1, 2, · · · electrons from this configuration, those on the right have 1, 2, · · · electrons more. Therefore, the former take on electrons easily (electron affinity) and the latter easily lose them (small ionization energy). Hence it happens that a pair of ions Na$^+$, Cl$^-$ is more stable than the neutral atoms Na, Cl. These pairs of ions attract each other with Coulomb forces. They would interpenetrate were it not for repulsive forces. Since we know nothing of the latter except that they suddenly become effective at a certain distance r, the immediate assumption is to try a law of the form $\dfrac{b}{r^n}$, where b and n are constants. Therefore the total elementary potential is

$$\phi_{kk'}(r) = \pm \frac{e^2}{r} + \frac{b_{kk'}}{r^n} \qquad (1)$$

where e is the charge of the ions. We now have to show that the ionic complexes which result from this law of force are precisely the observed ionic lattices and have their properties. It is clear at once that this formula can hold only for highly symmetrical lattices. In the cases of asymmetrically placed ions we must take into account the deformability of the ions (polarizability).

We now have the mathematical problem of calculating the total potential energy Φ of the lattice, but we encounter the difficulty that in summing over the lattice, the series

$$\phi(x, y, z) = S \sum_{l\ k} \frac{e_k}{|\mathbf{r}_k^l - \mathbf{r}|} \qquad (2)$$

converges very slowly. It is therefore necessary to transform it into a series which converges more rapidly. The simple arrangement of the terms according to the points of the lattice is then however, lost. We need special considerations because we have to calculate the value of the potential, at a

THE LATTICE THEORY OF RIGID BODIES 157

point of the lattice, of all the points of the lattice except the one at which the potential is calculated. This problem was first solved by Appell, but his work was not applied to physical problems or numerical computations. The first practical method was given by Madelung. He pointed out that it is very easy to calculate the energy of a linear, neutral series of points. For instance the points may be equidistant and with charges alternately $+e$ and $-e$. They can be numbered in both directions from any arbitrary point called 0.

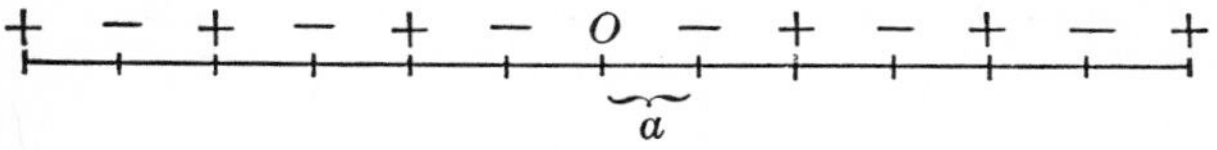

+ − + − + − O − + − + − +
  ⏟
  a

Fig. 22.

Then the potential at 0 of all points of the series, excepting 0, is evidently

$$\phi_0 = 2\frac{e^2}{a}\left(-\frac{1}{1} + \frac{1}{2} - \frac{1}{3} + \cdots (-1)^n \frac{1}{n} + \cdots\right)$$
$$= -2\frac{e^2}{a}\ln 2 \qquad (3)$$

and the energy of a series of N points is, except for the end corrections,

$$\frac{N}{2}\phi_0 = -\frac{Ne^2}{a}\ln 2.$$

A similar procedure is possible for any periodic series of points.

In the second place we shall consider a neutral lattice plane consisting of a parallel arrangement of linear series such as in Fig. 23. We calculate the potential at one of its points, for instance, the point 0, due to all its other points. This can be considered as the sum of the potential of a point series through 0 as calculated above, and the potentials at 0 of all other point series. Because the latter do not contain 0 they are easy to calculate (a sum of Bessel functions). If finally the whole crystal is considered as a sum of such lattice planes, the same procedure gives the total energy of the lattice.

But in many cases such a decomposition of the lattice in

neutral point series and lattice planes is not possible. Then a method devised by Ewald solves the problem. We shall briefly outline this method.

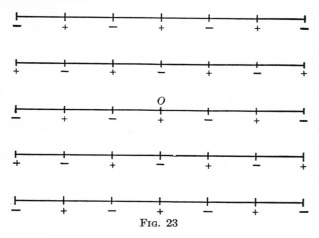

Fig. 23

Instead of the point-charges of the lattice we consider first a continuous periodic distribution of charges represented by a Fourier series without a constant term

$$\rho = \underset{l}{S'} \rho^l e^{i(\mathbf{q}^l \cdot \mathbf{r})}. \qquad (4)$$

Here $\mathbf{r}$ is the vector x, y, z and $\mathbf{q}^l = 2\pi(l_1\mathbf{b}_1 + l_2\mathbf{b}_2 + l_3\mathbf{b}_3)$ where $\mathbf{b}_1, \mathbf{b}_2, \mathbf{b}_3$ are the base vectors of the "reciprocal lattice." They are defined by the equation $\mathbf{a}_i \cdot \mathbf{b}_k = \delta_{ik}$ and have the values

$$\mathbf{b}_1 = \frac{1}{\Delta} \mathbf{a}_2 \times \mathbf{a}_3$$

$$\mathbf{b}_2 = \frac{1}{\Delta} \mathbf{a}_3 \times \mathbf{a}_1$$

$$\mathbf{b}_3 = \frac{1}{\Delta} \mathbf{a}_1 \times \mathbf{a}_2.$$

The coefficients ρ^l of the Fourier series are represented in the usual manner by the integral over a cell

$$\rho^l = \frac{1}{\Delta} \int \int \int \rho e^{-i(\mathbf{q}^l \cdot \mathbf{r})} \, dx \, dy \, dz. \qquad (5)$$

THE LATTICE THEORY OF RIGID BODIES

We shall also write ϕ as a Fourier series

$$\phi = S' c^l e^{i(\mathbf{q}^l \cdot \mathbf{r})}. \tag{6}$$

Poisson's differential equation gives

$$\nabla^2 \phi = -4\pi\rho$$

whence, by substituting and comparing the coefficients,

$$c^l = \frac{4\pi\rho^l}{|\mathbf{q}^l|^2}. \tag{7}$$

Now we pass to the limit when the continuous distribution of charges reduces to a series of point charges e_k, where $\sum_k e_k = 0$ at the extremities of the vectors $\mathbf{r}_k$. Then we have

$$\rho^l = \frac{1}{\Delta} \iiint \rho e^{-i(\mathbf{q}^l \cdot \mathbf{r})} \, dx \, dy \, dz$$

$$= \sum_k \frac{e_k}{\Delta} e^{-i(\mathbf{q}^l \cdot \mathbf{r}_k)} \tag{8}$$

and the potential becomes

$$\phi = \frac{4\pi}{\Delta} S'_l \sum_k \frac{e_k}{|\mathbf{q}^l|^2} e^{i\mathbf{q}^l \cdot (\mathbf{r} - \mathbf{r}_k)}. \tag{9}$$

We can write this as

$$\phi = \sum_k e_k \Psi(\mathbf{r} - \mathbf{r}_k) \tag{10}$$

where

$$\Psi = \frac{4\pi}{\Delta} S'_l \frac{e^{i\mathbf{q}^l \cdot \mathbf{r}}}{|\mathbf{q}^l|^2}. \tag{11}$$

These series also converge very slowly. Therefore, Ewald transforms them in the following way. By means of the identity

$$\frac{1}{a} = \int_0^\infty e^{-a\xi} \, d\xi$$

we obtain

$$\Psi = \frac{4\pi}{\Delta} \int_0^\infty S'_l e^{-|\mathbf{q}^l|^2 \xi + i\mathbf{q}^l \cdot \mathbf{r}} \, d\xi.$$

This integral is decomposed into two parts by an arbitrarily chosen number η intermediate between the limits of integration:

$$\Psi = \Psi_1 + \Psi_2;$$
$$\Psi_1 = \frac{4\pi}{\Delta} \int_\eta^\infty \sum_l S' e^{-|\mathbf{q}^l|^2 \xi + i\mathbf{q}^l \cdot \mathbf{r}} \, d\xi$$
$$\Psi_2 = \frac{4\pi}{\Delta} \int_0^\eta \sum_l S' e^{-|\mathbf{q}^l|^2 \xi + i\mathbf{q}^l \cdot \mathbf{r}} \, d\xi. \qquad (12)$$

The first expression can be calculated by elementary processes

$$\Psi_1 = \frac{4\pi}{\Delta} \sum_l S' \frac{e^{-|\mathbf{q}^l|^2 \eta + i\mathbf{q}^l \cdot \mathbf{r}}}{|\mathbf{q}^l|^2}. \qquad (13)$$

This expression resembles the former Fourier series for Ψ except that every term has an exponential factor which decreases very rapidly for increasing l_1, l_2, l_3, which improves the convergence.

The second part Ψ_2 can also be reduced to known functions. To do this Ewald used a formula in the theory of theta-functions to demonstrate the identity

$$\frac{4\pi}{\Delta} \sum_l S e^{-|\mathbf{q}^l|^2 \xi + i\mathbf{q}^l \cdot \mathbf{r}} = \frac{1}{2\sqrt{\pi \xi^3}} \sum_l S e^{-\frac{1}{4\xi}(\mathbf{r}^l - \mathbf{r})^2}.$$

We therefore obtain

$$\Psi_2 = \int_0^\eta \left\{ \frac{1}{2\sqrt{\pi \xi^3}} \sum_l S e^{-\frac{1}{4\xi}(\mathbf{r}^l - \mathbf{r})^2} - \frac{4\pi}{\Delta} \right\} d\xi.$$

Setting

$$\alpha = \frac{1}{2\sqrt{\xi}}, \qquad \epsilon = \frac{1}{2\sqrt{\eta}},$$

we get

$$\Psi_2 = \frac{2}{\sqrt{\pi}} \int_\epsilon^\infty \sum_l S e^{-\alpha^2 (\mathbf{r}^l - \mathbf{r})} \, d\alpha - \frac{\pi}{\Delta \epsilon^2}.$$

These integrals are known. They are related to Gauss's error function

$$F(x) = \frac{2}{\sqrt{\pi}} \int_0^x e^{-\alpha^2} \, d\alpha$$

THE LATTICE THEORY OF RIGID BODIES

or, more conveniently, with

$$G(x) = 1 - F(x) = \frac{2}{\sqrt{\pi}} \int_x^\infty e^{-\alpha^2}\, d\alpha.$$

Replacing η by ϵ in Ψ_1, we obtain finally

$$\Psi_1 = \frac{4\pi}{\Delta} S'_l \frac{e^{-\frac{1}{4\epsilon^2}|\mathbf{q}^l|^2 + i\mathbf{q}^l\cdot\mathbf{r}}}{|\mathbf{q}^l|^2}$$
$$\Psi_2 = S_l \frac{G(\epsilon|\mathbf{r}^l - \mathbf{r}|)}{|\mathbf{r}^l - \mathbf{r}|} - \frac{\pi}{\Delta\epsilon^2}. \tag{14}$$

The sum $\Psi_1 + \Psi_2$ is of course independent of the arbitrary dividing value ϵ. For $\epsilon = \infty$ we have $\Psi_2 = 0$ and Ψ_1 transformed into the original Fourier series. For $\epsilon = 0$ we have $\Psi_1 = 0$ and Ψ_2 becomes a divergent series, but the total potential ϕ takes its "Coulomb form" if the summation is made with a suitable arrangement of the terms

$$\phi = \sum_k e_k \Psi(\mathbf{r} - \mathbf{r}_k) = S \sum_{l\ k} \frac{e_k}{|\mathbf{r}_k^l - \mathbf{r}|}. \tag{15}$$

The smaller the value of ϵ, the better Ψ_1 converges. Ψ_2 converges better the greater the value of ϵ. By a suitable choice of ϵ we can make both series converge rapidly.

To calculate the value of the potential at a point k' of the base we have to subtract from ϕ the value of $\dfrac{e_{k'}}{|\mathbf{r} - \mathbf{r}_{k'}|}$. We thus obtain the "exciting potential,"

$$\phi_{k'}(\mathbf{r}) = e_{k'}\overline{\Psi}(\mathbf{r} - \mathbf{r}_{k'}) + \sum_k{}' e_k \Psi(\mathbf{r} - \mathbf{r}_k) \tag{16}$$

where
$$\overline{\Psi}(\mathbf{r}) = \Psi(\mathbf{r}) - \frac{1}{r}.$$

The subtraction of the term $\dfrac{1}{r}$ from Ψ can also be done by means of the intermediate value ϵ, placing

$$\frac{1}{r} = \frac{2}{\sqrt{\pi}} \int_0^\infty e^{-r^2\alpha^2}d\alpha = \frac{2}{\sqrt{\pi}} \int_0^\epsilon e^{-r^2\alpha^2}d\alpha + \frac{2}{\sqrt{\pi}} \int_\epsilon^\infty e^{-r^2\alpha^2}d\alpha.$$

The first integral which is to be subtracted from Ψ_1 has the value $\frac{1}{r} F(\epsilon r)$. The second is just equal to the term $l = 0$ of Ψ_2. Therefore

$$\overline{\Psi} = \overline{\Psi}_1 + \overline{\Psi}_2$$

$$\begin{cases} \overline{\Psi}_1 = \frac{4\pi}{\Delta} \underset{l}{S'} \frac{e^{-\frac{1}{4\epsilon^2}|\mathbf{q}^l|^2 + i\mathbf{q}^l \cdot \mathbf{r}}}{|\mathbf{q}^l|^2} - \frac{F(\epsilon r)}{r} \\ \overline{\Psi}_2 = \underset{l}{S'} \frac{G(\epsilon|\mathbf{r}^l - \mathbf{r}|)}{|\mathbf{r}^l - \mathbf{r}|} - \frac{\pi}{\Delta \epsilon^2}. \end{cases} \quad (17)$$

To calculate the potential of all points at a certain point k' we must set $\mathbf{r} = \mathbf{r}_{k'}$. We obtain

$$\phi_{k'} = \phi_{k'}(\mathbf{r}_{k'}) = e_k \overline{\Psi}(0) + \underset{k}{\Sigma'} e_k \Psi(\mathbf{r}_{kk'})$$

where

$$\begin{cases} \overline{\Psi}_1(0) = \frac{4\pi}{\Delta} \underset{l}{S'} \frac{e^{-\frac{1}{4\epsilon^2}|\mathbf{q}^l|^2}}{|\mathbf{q}^l|^2} - \frac{2\epsilon}{\sqrt{\pi}} \\ \overline{\Psi}_2(0) = \underset{l}{S'} \frac{G(\epsilon r^l)}{r^l} - \frac{\pi}{\Delta \epsilon^2}. \end{cases}$$

LECTURE 5

The energy of the rock-salt lattice — Repulsive forces — Derivation of the properties of salt crystals from the properties of inert gases.

As an example I shall take the familiar case of rock salt. I shall choose as the cell, not the elementary cube with the side a, but the rhombohedron defined by*

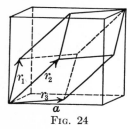

Fig. 24

$$\mathbf{a}_1 = \left(0,\ \frac{a}{2},\ \frac{a}{2}\right)$$
$$\mathbf{a}_2 = \left(\frac{a}{2},\ 0,\ \frac{a}{2}\right)$$
$$\mathbf{a}_3 = \left(\frac{a}{2},\ \frac{a}{2},\ 0\right)$$

whose volume is

$$\Delta = (\mathbf{a}_1 \times \mathbf{a}_2) \cdot \mathbf{a}_3 = 2\left(\frac{a}{2}\right)^3 = \delta^3$$

$$\delta = \frac{a}{2}\sqrt[3]{2}.$$

The base consists of the points

Na^+-Ions, $k = 1$, $\mathbf{r}_1 = (0, 0, 0)$, $e_1 = e$;

Cl^--Ions, $k = 2$, $\mathbf{r}_2 = \left(\frac{a}{2},\ \frac{a}{2},\ \frac{a}{2}\right)$, $e_2 = -e$.

The reciprocal lattice is given by:

$$\mathbf{b}_1 = \left(-\frac{1}{a},\ \frac{1}{a},\ \frac{1}{a}\right)$$
$$\mathbf{b}_2 = \left(\frac{1}{a},\ -\frac{1}{a},\ \frac{1}{a}\right)$$
$$\mathbf{b}_3 = \left(\frac{1}{a},\ \frac{1}{a},\ -\frac{1}{a}\right).$$

* $\mathbf{a}_1 = \left(0, \frac{a}{2}, \frac{a}{2}\right)$, etc. means the vector $\mathbf{a}_1$ having components $0, \frac{a}{2}, \frac{a}{2}$.
In Gibbs' notation $\mathbf{a}_1 = \mathbf{i}\,0 + \mathbf{j}\frac{a}{2} + \mathbf{k}\frac{a}{2}$.

We have

$$\mathbf{r}_{21} = \mathbf{r}_2 - \mathbf{r}_1 = \left(\frac{a}{2}, \frac{a}{2}, \frac{a}{2}\right)$$

$$\mathbf{r}^i = \left[\frac{a}{2}(l_2 + l_3), \frac{a}{2}(l_3 + l_1), \frac{a}{2}(l_1 + l_2)\right]$$

$$\mathbf{r}^i - \mathbf{r}_{21} = \left[\frac{a}{2}(l_2 + l_3 - 1), \frac{a}{2}(l_3 + l_1 - 1), \frac{a}{2}(l_1 + l_2 - 1)\right]$$

$$\mathbf{q}^i = \left[\frac{2\pi}{a}(-l_1 + l_2 + l_3), \frac{2\pi}{a}(l_1 - l_2 + l_3), \frac{2\pi}{a}(l_1 + l_2 - l_3)\right].$$

Instead of summing over l_1, l_2, l_3, new summation indices can be introduced by the transformations

$$\begin{array}{ll} -l_1 + l_2 + l_3 = l_1' & l_2 + l_3 = l_1'' \\ l_1 - l_2 + l_3 = l_2' & l_3 + l_1 = l_2'' \\ \underline{l_1 + l_2 - l_3 = l_3'} & \underline{l_1 + l_2 = l_3''} \\ l_1 + l_2 + l_3 = l_1' + l_2' + l_3' & 2(l_1 + l_2 + l_3) = l_1'' + l_2'' + l_3'' \end{array}$$

$$\begin{array}{l} l_2 + l_3 - 1 = l_1''' \\ l_3 + l_1 - 1 = l_2''' \\ \underline{l_1 + l_2 - 1 = l_3'''} \\ 2(l_1 + l_2 + l_3 - 1) - 1 = l_1''' + l_2''' + l_3''' \end{array}$$

and it is seen from the summation relations given above that l_1', l_2', l_3', can take all values; l_1'', l_2'', l_3'', only such the sum of which is even, and l_1''', l_2''', l_3''' only such the sum of which is odd. We therefore have

$$\phi_1 = -\phi_2 = e[\overline{\Psi}(0) - \Psi(\mathbf{r}_{21})],$$

$$\begin{cases} \overline{\Psi}_1(0) = \dfrac{4}{\pi a} {S'_l} \dfrac{e^{-\frac{\pi^2}{a^2\epsilon^2}(l_1^2 + l_2^2 + l_3^2)}}{l_1^2 + l_2^2 + l_3^2} - \dfrac{2\epsilon}{\sqrt{\pi}} \\[2ex] \overline{\Psi}_2(0) = \dfrac{2}{a} {S'_{\substack{l_1+l_2+l_3 \\ \text{even}}}} \dfrac{G\left(\dfrac{\epsilon a}{2}\sqrt{l_1^2 + l_2^2 + l_3^2}\right)}{\sqrt{l_1^2 + l_2^2 + l_3^2}} - \dfrac{\pi}{\Delta \epsilon^2} \end{cases}$$

THE LATTICE THEORY OF RIGID BODIES

$$\begin{cases} \Psi_1(\mathbf{r}_{21}) = \dfrac{4}{\pi a} \underset{l}{S'} \dfrac{e^{-\dfrac{\pi^2}{a^2\epsilon^2}(l_1^2+l_2^2+l_3^2)+i\pi(l_1+l_2+l_3)}}{l_1^2 + l_2^2 + l_3^2} \\[2ex] \Psi_2(\mathbf{r}_{21}) = \dfrac{2}{a} \underset{\substack{l_1+l_2+l_3\\ \text{odd}}}{S'} \dfrac{G\left(\dfrac{\epsilon a}{2}\sqrt{l_1^2+l_2^2+l_3^2}\right)}{\sqrt{l_1^2+l_2^2+l_3^2}} - \dfrac{\pi}{\Delta\epsilon^2} \end{cases}$$

and we have a certain control over the numerical calculation by computing twice with different values of ϵ.

The energy per cell is

$$\frac{\phi_0}{2} = \tfrac{1}{2}(e\phi_1 - e\phi_2) = e\phi_1 = e^2[\overline{\Psi}(0) - \Psi(\mathbf{r}_{21})].$$

By a simple transformation we obtain

$$\frac{\phi_0}{2} = \frac{e^2}{a}\left\{ \frac{8}{\pi} \underset{\substack{l_1+l_2+l_3\\ \text{odd}}}{S'} \frac{e^{-\dfrac{\pi^2}{a^2\epsilon^2}(l_1^2+l_2^2+l_3^2)}}{l_1^2+l_2^2+l_3^2} - \frac{2a\epsilon}{\sqrt{\pi}} \right.$$

$$\left. + 2\underset{l}{S'}(-1)^{(l_1+l_2+l_3)} \frac{G\left(\dfrac{a\epsilon}{2}\sqrt{l_1^2+l_2^2+l_3^2}\right)}{\sqrt{l_1^2+l_2^2+l_3^2}} \right..$$

Numerical computations give, for the NaCl type,

$$\frac{\phi_0}{2} = -\frac{e^2}{a} 3.495115 = -\frac{e^2}{\delta}\frac{\sqrt[3]{2}}{2} 3.495115.$$

The electrostatic energies of other lattice types (caesium chloride, zinc blende, Wurtzite, fluorspar, cuprite, rutile, anatase) are calculated in similar ways.

We can apply the law of force

$$\phi = -\frac{a}{r^m} + \frac{b}{r^n}, \quad \Phi_0 = -\frac{A}{\delta^m} + \frac{B}{\delta^n}$$

to the relation between energy and compressibility. We had (Lecture 3, Formulas (4) and (5))

$$\kappa = \frac{9\,\delta_0}{\left(\dfrac{d^2\overline{\Phi}_0}{d\delta^2}\right)_0} = \frac{9\,\delta_0^{m+3}}{A m(n-m)},$$

$$\Phi_0 = -\frac{A(n-m)}{n\delta_0^m} = -\frac{9\,\delta_0^3}{nm}\frac{1}{\kappa}.$$

The negative energy of a mole of the crystal will be called the "lattice energy" and denoted by U. If N is the number of molecules per mole, then

$$U = -N\Phi_0 = -\frac{N}{2}\phi_0 = N\left(\frac{\alpha e^2}{\delta} - \frac{\beta}{\delta^n}\right)$$

where α is the number computed above (for rock salt $\alpha = \frac{\sqrt[3]{2}}{2} 3.495115$).

We have, therefore, to set $m = 1$ and $A = \alpha e^2$. Then

$$\begin{cases} \kappa = \dfrac{9\,\delta_0^4}{\alpha e^2(n-1)}, \quad n = 1 + \dfrac{9}{\alpha e^2}\dfrac{\delta_0^4}{\kappa} \\ U = \left(1 - \dfrac{1}{n}\right)N\alpha\dfrac{e^2}{\delta_0}. \end{cases}$$

Therefore we can compute n, the exponent in the law of repulsion, and the lattice energy U, from δ_0 and κ. If M is the molecular weight

$$M = \delta^3 N \rho$$

then

$$n = 1 + \frac{7.701 \times 10^{-13}}{\alpha\kappa}\left(\frac{M}{\rho}\right)^{\frac{4}{3}}$$

and in thermal units

$$U = 279.1\,\alpha\left(1 - \frac{1}{n}\right)\left(\frac{\rho}{M}\right)^{\frac{1}{3}} \quad \text{kilocalories.}$$

For the crystals of the alkali halides, the first formula gives values of n between 7.8 and 9.8. Using a mean value $n = 9$ in the last formula we obtain

Type NaCl: $\qquad U = 545\left(\dfrac{\rho}{M}\right)^{\frac{1}{3}}$

Type CaF$_2$: $\qquad U = 1770\left(\dfrac{\rho}{M}\right)^{\frac{1}{3}}$

Type ZnS: $\qquad U = 2120\left(\dfrac{\rho}{M}\right)^{\frac{1}{3}}$

THE LATTICE THEORY OF RIGID BODIES 167

These formulas give reasonable approximations for the lattice energy. This is of course the case, because the major part of the energy is electrostatic, the repulsive forces giving only the fraction corresponding to $\frac{1}{n}$. The question of the energy content, therefore, does not depend on an accurate knowledge of the repulsive forces.

Since the ions of which the crystals under consideration are built have the configuration of the inert-gas atoms, it is reasonable to assume that they have approximately the same values of b and n. This conception has been developed further by Lennard-Jones and Taylor, who found the relative values of the constant b for ions of similar electron configuration (O^{--}, F^-, Ne, Na^+, Mg^{++} and S^{--}, Cl^-, A, K^+, Ca^{++}, etc.), from the ratio of their molal indices of refraction, which was possible because both of these quantities depend in a simple way on the radius of the ion. The absolute values of b were then found for the inert gases by the kinetic theory.

The determination of n is even more important than that of b. It was assumed to be the same for all the elements of each series, and its value was found from the properties of the inert gas in the series. For the neon series $n = 11$, for the argon series $n = 9$, for the krypton series $n = 10$, and for the xenon series $n = 11$. These values are in good agreement with the results of the lattice theory.

In this way the properties of salt crystals can be calculated absolutely from kinetic-theory data of the inert gases and the molal refraction of the ions. This very satisfactory result shows in a most beautiful way how distant branches of physics are interdependent. Had I known of this beautiful work at the beginning of the lectures, I should have built up the whole theory of ion lattices on this basis.

LECTURE 6

Experimental determination of the lattice energy by means of cyclic processes — The electron affinity of halogens — Heat of dissociation of salt molecules — Theory of molecular structure.

Unfortunately, the lattice energy is not a directly measurable quantity, but it can be related to measurable quantities. We may imagine a cyclic process as shown in the following figure:

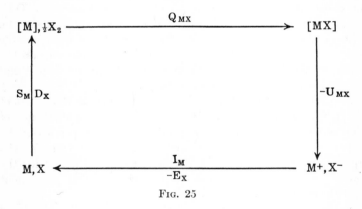

Fig. 25

Let us start at the upper left-hand corner with a mole of a solid metal [M] and a half mole of the gaseous, diatomic halogen $\frac{1}{2} X_2$. These unite to form the solid salt $[MX]$ with the liberation of the directly measurable heat of chemical reaction Q_{MX}. We now imagine the salt dispersed into an ionic gas M^+ and X^- with absorption of the lattice energy $(-U_{MX})$. We remove one electron from the halogen ion, by which the energy $(-E_X)$ is liberated, where E_X is a measure of the electron affinity. At the same time we allow the electron to become attached to the metal ion by the liberation of the work of ionization I_M. We now have a gas composed of the neutral atoms M and X. We allow the metallic atoms to solidify with

THE LATTICE THEORY OF RIGID BODIES 169

liberation of the heat of sublimation S_M. The halogen atoms unite to form molecules with liberation of the heat of dissociation D_X. We have then returned to the initial state $[M]$, $\frac{1}{2} X_2$. There is therefore no change in the energy of the system for the whole process, and

$$Q_{MX} - U_{MX} + (I_M + S_M) + (D_X - E_X) = 0. \qquad (1)$$

In this equation the following quantities are experimentally determinable:

Q_{MX}, by calorimetric measurements;

S_M, by direct measurement or by observing the pressure of sublimation as a function of the temperature (using Clausius-Clapeyron's formula);

D_X, by measuring the dissociation as a function of the temperature (using van t' Hoff's equation);

I_M, by the method of electron impact of Franck and Hertz controlled by optical measurements ($I_M = h\nu_\infty$, where h is Planck's constant and ν_∞ the frequency of the short wave limit of the principal series of the spectrum of the metal vapor).

Except for U_{MX}, the quantity to be determined, the only unknown is the electron affinity of the halogens. We must therefore make more measurements in order to eliminate E_X. Foote and Mohler in this country, and Knipping in Germany have undertaken experiments to decompose the halogen acids by electron impact. The results of these experiments have been rendered very doubtful by more recent investigations, but, in the absence of more reliable information, we shall make use of them here. We denote the work necessary to produce this decomposition by U_{HX}, because it is analogous to the lattice energy. We then obtain by a similar cyclic process the relation

$$Q_{HX} - U_{HX} + (I_H + D_H) + (D_X - E_X) = 0. \qquad (2)$$

where

Q_{HX} is the heat of combination measured calorimetrically,
D_H, the heat of dissociation of the hydrogen molecule, which is known approximately, and

I_H is the work of ionization of the hydrogen atom. It is not directly measurable, but can be calculated very accurately by Bohr's theory.

If we subtract this equation from that above, the terms $(D_X - E_X)$, corresponding to the halogen atom, cancel, and we obtain

$$U_{MX} = Q_{MX} + (I_M + S_M) - Q_{HX} + U_{HX} - (I_H + D_H). \quad (3)$$

Under these assumptions we get,

	U Observed	U Calculated
NaCl	183 k. cal.	182 k. cal.
NaBr	170	171
NaI	159	158
KCl	160	162
KBr	154	155
KI	144	144
RbCl	161	155
RbBr	151	148
RbI	141	138

At present, the "calculated" values seem to be much more reliable than the "observed." In any case the sequence of the values agrees throughout. New experimental data would be very desirable.

We may now estimate the electron affinity of the halogen atoms, and find:

$$E_{Cl} = 86 \text{ k. cal.}$$
$$E_{Br} = 86$$
$$E_I = 79$$

The values for chlorine and bromine should not be the same, but an error of ±5 k. cal., or even more, is explainable by the inaccuracy of the data. It would certainly be important to have an independent investigation of these quantities. Franck has had the following idea: if we have, in a halogen gas at a high temperature, a considerable number of negative ions X^-, we should observe a continuous spectrum with a sharp limit on the long-wave side. This spectrum arises, in the case of absorption, for instance, from the removal of the extra electron by the action of light. The smallest frequency ν_{min} which is able to do this corresponds to the electron affinity, according to the formula

$$E = h\nu_{min}.$$

Franck thought he had found this limit in the emission spectrum of iodine, given by Steubing, and of bromine, given by Eder and Valenta. More accurate investigations, to which Steubing, von Angerer, Gerlach, and Oldenberg have contributed, have, however, shown that these emission spectra were a kind of band spectrum of another origin. Recently Angerer and Müller believed they had found Franck's spectrum in absorption, and gave the values,

$$E_F = 103 \text{ k. cal}$$
$$E_{Cl} = 100$$
$$E_{Br} = 91$$
$$E_I = 79.$$

Introducing these numbers, the values for U-observed given above are slightly changed. A new discussion of these quantities, using the latest results of experimental research, should be undertaken at once.

An estimate of the electron affinity of the sulphur atom, that is, the work necessary to remove from the S^{--}-ion its two electrons, has been made by Gerlach and myself using the lattice energies of crystals of galena (PbS) and zinc blende (ZnS). We found

$$E_S = 45 \text{ k. cal.}$$

The results of the corresponding calculation for oxygen are even less reliable.

If we knew the energies of the molecules of the vapor of the salt MX, that is, the work necessary to separate MX into the ions M^+, X^-, we should have another method for determining the energy of the lattice, by the following cyclical process (Fig. 26):

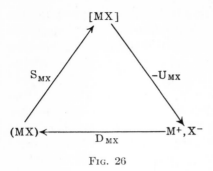

Fig. 26

Let [MX] be the solid salt. It is separated into the ions M^+, X^-, by giving to it the lattice energy U_{MX}. These ions unite to form the molecule MX, liberating the heat of dissociation D_{MX}. The molecules then condense to the solid salt [MX], liberating the heat of sublimation S_{MX}. We therefore have

$$U_{MX} = S_{MX} + D_{MX}. \tag{4}$$

S_{MX} has recently been very accurately measured by von Wartenberg and his students. D_{MX} is, however, not known. Heisenberg and I have therefore found a way of calculating D_{MX} theoretically. It is necessary, as I have already said, to consider not only the electrostatic forces, but also the polarizability of the ions. In diatomic molecules, every ion is acted on by unilateral forces, and therefore deformed. The potential energy of the molecules of the vapor of the salt is

$$\Phi = -\frac{e^2}{r} + \frac{b}{r^n} + \frac{p_1 e - p_2 e}{r^2} + \frac{p_1^2}{2\alpha_1} + \frac{p_2^2}{2\alpha_2} \tag{5}$$

where p_1, p_2 are the electric moments of the two ions, α_1 and α_2 their polarizabilities. The conditions of equilibrium for p_1 and p_2 give

$$\frac{\partial \Phi}{\partial p_1} = \frac{e}{r^2} + \frac{p_1}{\alpha_1} = 0 \qquad \frac{\partial \Phi}{\partial p_2} = -\frac{e}{r^2} + \frac{p_2}{\alpha_2} = 0,$$

whence

$$p_1 = -\frac{\alpha_1 e}{r^2} \qquad p_2 = \frac{\alpha_2 e}{r^2}$$

and, therefore,

$$\Phi = -\frac{e^2}{r} + \frac{b}{r^n} - \frac{e^2}{2}(\alpha_1 + \alpha_2)\frac{1}{r^4}. \tag{6}$$

The values of n (e.g., $n = 9$) and b, are taken from the lattice theory. α_1 and α_2 are determinable from the dielectric constant ϵ of the salt or its solution, or might be found from the more accurately measurable index of refraction n for long waves, when Maxwell's law $n^2 = \epsilon$ holds. These measurements and calculations have been made by Heydweiller, Wasastjerna, Fajans and Joos, and also by Heisenberg and myself. For simplicity let us consider the vapor of the salt. The electric moment P per unit of volume (N molecules) produced by the field E is evidently

$$P = N(\alpha_1 + \alpha_2)E,$$

where it is supposed that all ions are deformed independently of one another. According to Maxwell's theory,

$$4\pi P = (\epsilon - 1)E = (n^2 - 1)E.$$

Therefore

$$\alpha_1 + \alpha_2 = \frac{n^2 - 1}{4\pi N}. \qquad (7)$$

If we do not have a vapor, but a solution or a solid salt, then we must also consider the close packing of the ions. This is done roughly by introducing instead of $(n^2 - 1)$, the expression of Lorentz-Lorenz

$$3\frac{n^2 - 1}{n^2 + 2}.$$

The condition for equilibrium is

$$\frac{d\Phi}{dr} = \frac{e^2}{r^2} - \frac{nb}{r^{n+1}} + 2e^2(\alpha_1 + \alpha_2)\frac{1}{r^5} = 0. \qquad (8)$$

We now have all the data necessary to calculate, from this equation, the equilibrium distance r_0 and hence the energy Φ_0 of the molecule. This value of Φ_0 multiplied by the number of molecules in a mole is equal to $-D_{MX}$. If we introduce these numerical values in the equation $U_{MX} = S_{MX} + D_{MX}$, we obtain values for the lattice energy which agree with those calculated directly. If we do not consider the polarization the agreement is not so close.

These simple considerations in a theory of the molecule have been carried further by my associates. Heisenberg first investigated triatomic molecules, of the type of water vapor, H_2O, for instance, and showed that the symmetrical arrangement

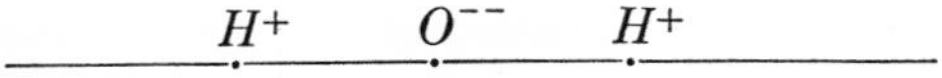

$$H^+ \quad O^{--} \quad H^+$$

Fig. 27

is of course the only stable form for a non-polarizable central atom, but for a polarizable central atom there exists a more stable linear form with asymmetrically placed ions:

$$H^+ \quad O^{--} \qquad H^+$$

Fig. 28

Hund has shown that even this form is not the most stable, but rather an isosceles triangle as shown in Fig. 29.

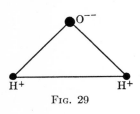

Fig. 29

Quantitative calculations of the energy values are not immediately possible, because the H^+-ions, the simple H-nuclei, are inside the O^{--}-ion. Therefore Hund has developed a new method which no longer postulates *a priori* the energy-function, but determines it empirically from all the known data, in particular those obtained from band spectra. This method has succeeded for the halogen acids, and gives information about the arrangement of the ions in the molecules H_2O, NH_3, CH_4, etc., and also the molecular distances and energies.

Other of my associates (Hund, Kornfeld, Rolan) have calculated the proper frequencies of such ions, for instance of the carbonate ion $CO_3^=$, and have compared them with the measurements of Schaefer and his pupils of the infra-red absorption bands.

The postulate that the cohesive forces of the lattice are of electrostatic origin has certainly been justified. It has proved useful in chemistry, quite independently of its application to the

THE LATTICE THEORY OF RIGID BODIES 175

lattice theory. The cyclic processes given above show clearly that the heats of reaction of chemical processes which can be measured calorimetrically are complicated sums of elementary quantities, which characterize on the one hand the individual ions, and on the other the arrangement of the ions. It is clear that these elementary quantities follow more simple laws than the resulting heats of reaction. This has been especially recognized by Grimm, who found that the relative lattice energies determined empirically are related to the atomic structure and to the periodic system of the elements in a way that can be easily interpreted. He was able to explain and systematize a very large number of inorganic chemical data. To these belong the property of many solid compounds of absorbing water or ammonia in considerable amounts. According to Biltz and Grimm this can be explained by an estimate of the lattice energy.

LECTURE 7

Chemical crystallography — Coördination lattices — Hund's theory of lattice types — Molecule, radical and layer lattices.

The question why some ions build a certain type of lattice and others build another type must now be considered. This is the basic problem of *chemical crystallography*. Hund has recently given a theoretical contribution from the point of view of the electrostatic nature of the forces between the ions. He tries to explain the appearance of the different types of lattice from purely energetical considerations. That structure is the most stable which, for certain given ions, leads to the greatest lattice energy ($U = -\Phi$) or to the lowest energy level. The ions are to be defined in terms of the least possible number of constants. If we are concerned with lattices in which ions of the same kind have the same surroundings, the so-called "coördination lattice" (after Kossel), the ions will only be distinguished by the constants b and n of the repulsive term br^{-n} in the potential energy. But since b can be eliminated by the condition of equilibrium, that is, can be reduced to the lattice constant, there remains only the exponent n as a constant of the ions in comparing coördination lattices. In more general lattices which have molecules as structural units, the deformability α enters.

The coördination lattices can be classified according to purely geometrical properties, that is, by the number of neighbors of each ion. From the chemical standpoint this is the "coördination number" of Werner. The following cases are possible:

(1) 12 neighbors in the directions of the face diagonals of a cube.

(2) 8 neighbors in the directions of the space diagonals of a cube.

THE LATTICE THEORY OF RIGID BODIES

(3) 6 neighbors in the directions of the sides of a cube.
(4) 4 neighbors in the directions of the space diagonals of a cube (one to a diagonal).
(5) 3 neighbors forming a triangle.
(6) 2 neighbors in a straight line.

For compounds of the form XY the following cases are possible:

Form XY:

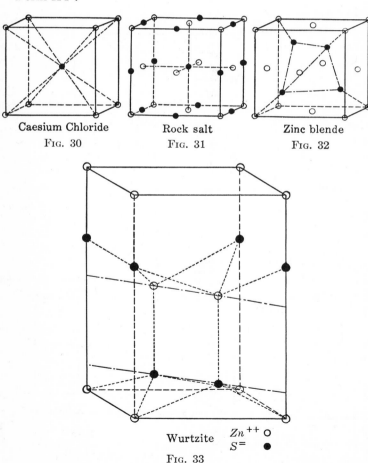

Caesium Chloride
Fig. 30

Rock salt
Fig. 31

Zinc blende
Fig. 32

Wurtzite $\quad Zn^{++}$ ○
$\quad\quad\quad\quad\;\; S^{=}$ ●
Fig. 33

178 PROBLEMS OF ATOMIC DYNAMICS

Coördination number 12   not possible
"   "   8   type of cæsium chloride
"   "   6   type of rock salt
"   "   4   type of zinc blende, Wurtzite.
"   "   3   not possible.

In compounds of the form XY_2, every X-ion has twice as many neighbors as the Y-ions. The following arrangements are obtained:

Form XY_2:

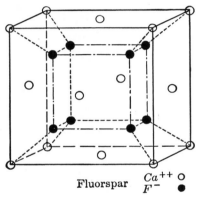

Fluorspar Ca^{++} ○
               F^- ●

Fig. 34

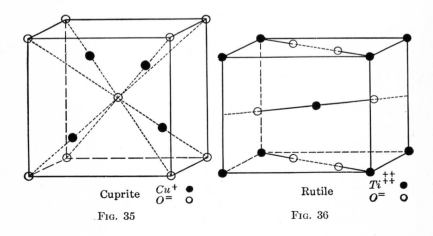

Cuprite Cu^+ ●
         $O^=$ ○

Fig. 35

Rutile Ti^{++}_{++} ●
       $O^=$ ○

Fig. 36

THE LATTICE THEORY OF RIGID BODIES

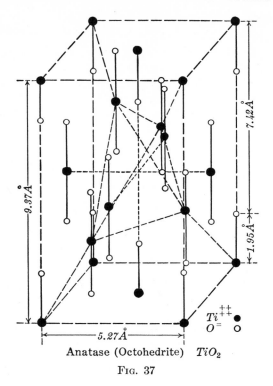

Anatase (Octohedrite) TiO_2

FIG. 37

Coördination numbers 8, 4 type of fluorspar
      "   "   4, 2 type of cuprite
      "   "   6, 3 cannot be rigorously constructed. Approximately realized in the two types of TiO_2, anatase and rutile.

These types will now be compared with respect to their energy stability. Let the lattice energy per molecule for a compound XY be

$$U = -\frac{Z^2 \alpha e^2}{r} + \frac{b}{r^n}.$$

Here r is the smallest distance between ions of different kinds; α, Madelung's constant of the lattice (per unit charge on the ion) which we have calculated; e, the charge of the electron; and Z, the valence of the ion.

If n is large, the repulsive force br^{-n} acts as if the ions were rigid spheres. Then neighboring ions of different kinds come in contact at a certain distance r which is independent of the type of lattice. The energy of the lattice is therefore directly measured by the value of α. For the rock-salt type we found above

$$\frac{\phi_0}{2} = -\frac{e^2}{a}3.495\cdots$$

where a is the side of the elementary cube, that is, $a = 2r$. Therefore, here

$$Z = 1, \alpha = \tfrac{1}{2}3.495 = 1.748.$$

If we make the calculation for other types, we find:

Coördination number 8   type CsCl $\alpha = 1.762$
"   "   6   "   NaCl $\alpha = 1.748$
"   "   4   "   ZnS $\alpha = 1.639$

We should, therefore, for large n, expect that the type of lattice in which every ion has the greatest number of neighbors, i.e., CsCl, is the most stable. Then follows NaCl, and then ZnS, where the types zinc blende and Wurtzite have almost the same value of α. For a smaller value of n, the repulsive force has a greater effect, and its influence is greatest in the case of the greatest number of neighbors. Therefore, the order of stability of the three types will be changed. Hund has calculated, partly from the tables of Jones and Ingram, the lattice sums of the repulsive potential for values of n from 4 to 36, varying also the value of b for the different ions. His results are:

For $n > 35$, the CsCl type is the most stable.
     $35 > n > 6$, " NaCl " " "
     $n < 6$, " ZnS " " "

This agrees qualitatively with the results of observation. The values of n are found from the compressibility, or if this is not known, they can be estimated from the lattice energy determined by the cyclic process. We find for the salts CsCl, CsBr, CsI, which crystallize in the CsCl type, that in all cases n is greater than for the other alkali halides. The quantitative agreement, however, is not good, for in these salts the limit $n = 10$ holds instead of $n = 36$. For ZnS the value of n is extremely small ($n = 4$), as would be expected from the theory. Hund has discussed in this way the entire known material with the result that his explanation is seen to give at least the salient features of the observed phenomena.

This theory also holds for compounds of the type XY_2. Here the calculation of the constant α gives the order:

Coördination number: 8, 4, type fluorspar; $\alpha = 5.039$
" " 6, 3, type rutile (anatase) $= 4.81$
" " 4, 2, type cuprite $= 4.115$

For large n, the fluorspar type must be the most stable. With decreasing n, it is found that near $n = 7$ the rutile type is more stable, but the stability of the cuprite type cannot be explained. In fact, in the fluorides of the alkalies and the oxides of the tetravalent elements (inert-gas type ions), the fluorspar type is most common, and only in the case of very small ions does the rutile (or anatase) type appear. It is not to be expected, however, that the simple spherically symmetrical ion should suffice to explain all cases, especially where the ions are not of the inert-gas type (as Cu^+).

If we also consider the polarizability, which has been neglected so far, quite new types of lattices become possible: molecule lattices, such as HCl; radical lattices, such as $CaCO_3$; and layer lattices, in which groups of two or three lattice planes are closely associated, as in the type CdI_2. In this lattice the less polarizable cadmium ions are trigonally distributed in a plane which lies between two planes trigonally occupied by iodine ions (Fig. 38).

These lattices can now be investigated in a manner similar to that employed in the case of the ion and radical lattices, calculating their stability in terms of the polarizability α. The result found is that for small α the coördination lattices are the most stable. For large α, however, the molecule, radical or layer lattices may become more stable, because the deformation energy in the molecules, radicals, or layers, becomes greater than the decrease in the lattice energy due to the loosening of the packing. These energy relations depend on the ratio of the constant α, which has the dimensions of a volume

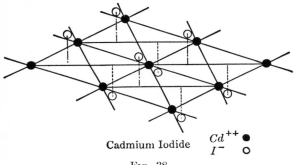

Cadmium Iodide Cd^{++} ●
 I^{-} ○

Fig. 38

to the cube of the ionic distance r. If one of the ions has a small volume, r is small, and therefore α is large with respect to r. A molecular or layer lattice is, therefore, to be expected. In the case of the hydrogen compounds, HCl, for example, where one of the ions is reduced to the hydrogen nucleus, the lattices are of the molecular type. In CdI_2, where cadmium carries a double positive charge and is, therefore, much smaller than the negatively charged iodine, we have a layer lattice. Many other cases can be similarly explained.

LECTURE 8

Physical mineralogy — The parameters of asymmetrical lattices — The molecule lattice of hydrochloric acid — Bragg's calculation of the rhombohedral angle of calcite — Rutile and anatase — Influence of the polarizability on elastic and electric constants — The breaking stress of rock salt.

In strictly coördination lattices, the form of the cell and the position of the lattice points are determined by properties of symmetry. In lattices which only approximately satisfy the coördination conditions (similar neighborhood for similar particles), and especially in molecule, layer and radical lattices, certain parameters enter, that is, elements which are not determined by symmetry. Therefore, we have the problem of calculating these parameters theoretically, which is the basic problem of crystallography.

With our present knowledge of molecular forces, we can only hope to find an answer if we can assume that the major part of the forces concerned are of electrostatic origin. Molecular lattices must in general be left out of consideration. The only attempt at a quantitative estimate has been made by Kornfeld and myself for the halogen acids in the solid state. We calculated the energy of the lattice on the assumption that the molecules of the HCl type exerted forces similar to those of an electric dipole. In this case the energy of the lattice is equal to the heat of sublimation. The moment of the dipole, or its length, found by dividing the moment by e, enters as a parameter. According to our formula it is possible to calculate this length from the heat of sublimation if the structure of the lattice is known. Assuming a simple regular lattice, we found for the length of the dipole a value of the same order of magnitude as that found by quite a different method (variation in the dielectric constant of the gas with temperature, according to Debye). Later, the true lattice was determined by X-ray observations by Simon and Simson, and found to differ from our assumptions, but a repetition of our calculations on the basis of this lattice gave no better agreement for the length of the dipole. Due to the close packing of molecules in the lattice, it is not to be expected that their reactions can be readily explained in terms of fixed dipoles.

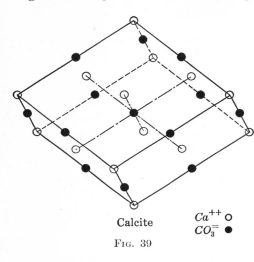

Calcite
FIG. 39

Ca^{++} ○
$CO_3^=$ ●

For lattices formed of radical ions greater success may be expected, because here the forces are the attractions between free charges. A problem of this type was attacked by Bragg. He investigated calcite, $CaCO_3$, and calculated a geometric parameter, the angle of the rhombohedral cell. In this crystal the radical ion CO_3^{--} has a double negative charge, and the metal ion Ca^{++}, a double positive charge. The CO_3^{--}-ion has the form of an equilateral triangle, with the C-atom at the center and the O-atoms at the vertices, an extremely stable system. Its cohesion is due to the fact that the C-atom has given up its four outer electrons and that these,

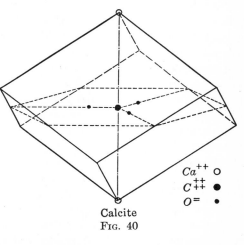

Calcite
FIG. 40

Ca^{++} ○
C^{++}_{++} ●
$O^=$ •

plus two foreign electrons, give the three O-atoms the stable

THE LATTICE THEORY OF RIGID BODIES 185

inert-gas form, and Bragg, therefore, assumed it to be rigid. Its dimensions have been determined by X-ray measurements. The calcite lattice can be described as a modified rock-salt lattice, but since one kind of ion is a triangle and not spherically symmetrical, the cell is not a cube, but a rhombohedron derived from the cube by a deformation in the direction of the space diagonal orthogonal to the planes of the triangles. Bragg assumed that the distances between neighboring ions CO_3^{--} — Ca^{++}, as well as the size of the triangles CO_3^{--}, are fixed. Then the height of the rhombohedron, or what amounts to the same, its angle, is continuously variable. This parameter was calculated by Bragg so as to minimize the electrostatic energy of the crystal. The angle which he found differs from the measured angle by only a few degrees. This error is due to the fact that the effective size of the CO_3^{--}-triangle is not given accurately by X-ray measurements. Bragg, therefore, chose this parameter so that the calculated and measured angle should agree.

To avoid this disadvantage I have tried to determine the parameters of lattices which are nearly coördination lattices. I chose the two modifications of the lattices of TiO_2, rutile and anatase. These are tetragonal lattices the coördination numbers of which are 6, 3. Every Ti^{++++}-ion has six O^{--} neighbors which tend to occupy the vertices of an octahedron. Every O^{--}-ion has three Ti^{++++} neighbors which tend to form an equilateral triangle. As a lattice which would allow both of these tendencies to be fulfilled does not exist, there must be a compromise. Keeping the distance of two neighbors Ti^{++++}, O^{--} fixed, the rutile and anatase lattices may still be continuously deformed.

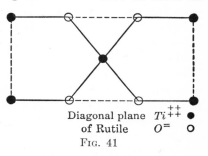

Diagonal plane Ti^{++} ●
of Rutile $O^{=}$ ○
Fig. 41

By varying the ratio of the height of the tetragonal cell to the side of its square base (Fig. 41), we pass continuously from the limiting case in which the O^{--}-ions occupy the vertices of an octahedron to that where the Ti^{++++}-ions form an equilateral triangle. For these two limiting cases the geometry of the lattices (Figs. 36, 37) gives:

For rutile $\gamma_{Ti} = 0.293$
$\gamma_0 = 0.333$
For anatase $\gamma_{Ti} = 0.250$
$\gamma_0 = 0.167$

The interval between these limits is relatively small, and therefore, a quasi-coördination compromise lattice is possible.

The true value of γ is now calculated by minimizing the electrostatic energy, neglecting small changes in the distances. We find for rutile $\gamma = 0.315$, while measurements on the crystals PbO_2, SnO_2, MgF_2, TiO_2, MnO_2 give values of γ between 0.303 and 0.315. The deviation shows dependence on the sizes of the ions, which can be explained theoretically. For anatase the calculated value of γ is 0.198, while observation gives 0.206.

The lattice energy per molecule is almost the same in the two cases. The constant α is 4.82 for rutile, and 4.80 for anatase, which explains the appearance of both forms in nature. In the chemical calculations given above the mean value, $\alpha = 4.81$, was used. These calculations were made by Bollnow.

We have seen from these considerations the rôle played by electrostatic forces in the structures of crystals. An approximate determination of the value of e, or at least of the valence of an ion, can be made from measurements of crystal properties, but the electrostatic theory is limited where the specific properties of the elements come into account. A consideration of the polarizability leads a trifle further, as has been shown by a few examples. Heckmann has carefully examined this question, and reaches the conclusion that, of the more accurately measurable properties of crystals, such as the elasticity, piezoelectricity, and dielectric properties, only a few can be quantitatively calculated, namely, those for which the deformability of ions does not play an important part. For regular crystals these are the two constants of elasticity, A and B, which appear in the relations between extension and normal pressure.

$$-X_x = A x_x + B(y_y + z_z).$$

THE LATTICE THEORY OF RIGID BODIES

These constants have been calculated for a few types of crystals from the ion charges and the lattice constants, and the results are in good agreement with experiment. Conversely, an approximate determination of e can be made from elastic measurements. For reasons already given, the theory fails to give good results for the piezoelectric and dielectric constants. The mechanism of these processes is too complicated to be explained by a linear polarizability of the ions.

In closing these considerations, I should like to give a verification of one of the bases of this theory, that is, of the assumption first formulated by Nernst that, in ion lattices, the mechanical cohesion of the rigid body and the chemical attraction of its particles are one and the same. Both appear as the electrostatic attraction between the charges of the ions. The proof is given by measurements of the breaking stress of crystals.

It is easy to calculate the breaking stress of rock salt if the law of force is known. To do this we consider the cubic rock-salt lattice with side a_0 to be deformed by a uniform tension normal to a cube face into a tetragonal lattice, the cell of which has a square base of side a and a height h. Then we can calculate the lattice energy $\Phi(a, h)$, as a function of a and h for the assumed law of force,

$$\phi = \pm \frac{e^2}{r} + \frac{b}{r^n}.$$

Since the transverse contraction is a consequence of the elongation, a is a function of h, and is found from the equation

$$\frac{\partial \Phi(a, h)}{\partial a} = 0.$$

If we introduce this function $a(h)$ in Φ, it becomes dependent only on h. Then the force necessary to produce the elongation is

$$K = -\frac{d\Phi}{dh}.$$

As h increases from a_0, that is, the side of the cube of the undeformed lattice, K first increases, reaches a maximum K_z, and then decreases very rapidly. K_z is the breaking stress.

Measurements of this quantity for rock salt were made many years ago by Voigt and Sella, and the numerical calculations published by Zwicky give values for K_z which are about 400 times greater than the measured breaking stress. The experiments, however, did not give the true "atomic" breaking stress, but a value influenced by surface effects, probably small cracks. This has been shown by Joffé and his associates by breaking the crystal under water, when the surface is freed of small cracks and irregularities by the action of the solvent. The formation of a crack on the surface is prevented, and the breaking stress is increased to about $\frac{3}{4}$ of the theoretical value. It is unimportant for us in these experiments whether we measure the breaking stress of the crystal in its original state, or if slip processes first produce a "strengthening" by eliminating the irregularities. It is only necessary to show that the breaking stress calculated from the electrostatic attraction of the ions really appears under suitable conditions. If this is true, the mechanical cohesion is identical with the forces introduced in the calculation of the lattice energy as "chemical" forces. Joffé's experiment is evidently a case where the "chemical affinities" are broken by mechanical means.

LECTURE 9

Crystal optics — Refraction and double refraction — Optical activity — Thermodynamics — Quantum theory of specific heats — Distribution of frequencies in phase space.

Many new developments in crystal optics have been made through the lattice theory. I shall only speak of the most essential points. Maxwell's electromagnetic theory of light gives the optical properties of substances, considered to be continuous, in terms of a few parameters, namely, the principal indices of refraction and the rotatory power. The problem of the atomic theory is not only to reduce these parameters to properties of atoms, but also to explain why optical phenomena can be described by Maxwell's continuum theory in spite of the atomic structure of crystals. After many previous efforts by other scientists (Rayleigh, Lorentz, Planck and others), this problem has been solved completely by Ewald through a thorough investigation of electromagnetic waves in dipole lattices. The mathematical methods are quite similar to those we have used above for the calculation of the electrostatic lattice potentials.

Ewald was able to show that a considerable part of the double refraction of non-cubic crystals is due to the structure of the lattice and not to the anisotropy of the atoms. The first quantitative results were obtained by Bragg, who calculated the double refraction of crystals like calcite and aragonite (both are $CaCO_3$) as a property of their structure alone. Ewald has developed his theory in another direction as well, applying it to X-rays, which are distinguished from light in that their wave-length is not large compared with the lattice constants. The results obtained have recently been confirmed by observations of the refraction of X-rays by Stenström, Hjalmar, Bragg, Siegbahn, Davis, Burger and Nardroff.

If we consider the influence of the finite size of the lattice constant with respect to the wave-length, we can calculate the rotatory properties of crystals. Earlier explanations had only described the phenomena, or made assumptions about the motions of electrons in the atom which are certainly not realized in nature. Independently, and almost at the same time, Oseen and myself found that this effect could be calculated without any new hypothesis. If the wave-length is no longer considered as infinitely large, but the ratio of the lattice constant to the wave-length taken account of to a first approximation, we obtain all the fine effects in crystals, described in

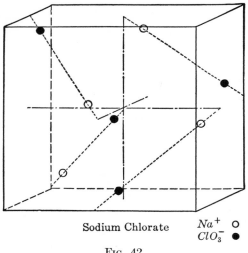

Sodium Chlorate Na^+ ○
ClO_3^- ●

Fig. 42

the formal theories of Drude, Voigt, Pockels, and others. We can go further and inquire which part of the optical activity is a property of the atoms and which part a property of the lattice. To answer this question, my associate Hermann has completely calculated the crystals of $NaBrO_3$ and $NaClO_3$. These crystals are cubic, and, therefore, not doubly refracting in the ordinary sense, but have the property of rotating the plane of polarization. As their ions in solutions are not optically active, this must be a property of the ion structure. Indeed, the structure determined by X-ray methods shows the characteristics necessary to produce rotation, that is, a screwlike arrangement of the ions. Considering the ions to be isotropic dipoles, a calculation of the rotatory power gives results which are in fair agreement with observed values.

In conclusion, let us return to the formal lattice theory and consider briefly the effects related to changes of temperature.

The modern theory of these processes started with the celebrated work of Einstein on the specific heat of rigid bodies, in which he first applied the quantum theory to such systems. Classical statistical mechanics states that, in thermal equilibrium, the mean of the kinetic energy of every degree of freedom has the same value, $\frac{kT}{2}$. The factor of proportionality k is called Boltzmann's constant. A crystal consisting of N atoms has $3N$ degrees of freedom, and, therefore, a mean kinetic energy of $\frac{3}{2} NkT$. If the amplitudes of the electron oscillations are small enough, the motion may be considered as a harmonic oscillation, and the mean of the potential energy is, by a well-known theorem, equal to the mean of the kinetic energy. The total thermal energy, E, is given by

$$E = 3NkT,$$

and the specific heat, C, is

$$C = 3Nk.$$

If a mole is considered,

$$Nk = R,$$

where R is the ideal gas constant, approximately 2 calories, and the molal specific heat is $c = 3R = 6$ calories approximately. This is the familiar law of Dulong and Petit, but, as is well known, it does not hold for low temperatures.

Einstein first recognized the reason for this deviation in the departure from the classical laws demanded by the quantum theory. Planck had found that the mean energy of a harmonic oscillator of frequency ν is not kT, but $kT\, P\left(\frac{h\nu}{kT}\right)$, where

Planck's function $P(x) = \dfrac{x}{e^x - 1}$ converges to 1 for small x, (large $\dfrac{kT}{h\nu}$), but for large x, (small $\dfrac{kT}{h\nu}$), decreases rapidly to zero. Planck succeeded in explaining this result by postulating energy quanta $h\nu$ (see the first series of lectures). Einstein applied Planck's hypothesis to crystals and obtained for the thermal energy

$$E = 3\,kT\,P\!\left(\frac{h\nu}{kT}\right), \tag{1}$$

a formula which gives a qualitative explanation of the dependence of specific heat on temperature.

Many publications by Debye, Kármán, myself, and others followed this, all attempting to replace the system of resonators of the same frequency ν, by a multiplicity of different resonators the frequency distribution of which coincided as far as possible with that of a real crystal. If there are $z(\nu)d\nu$ resonators the frequencies of which are in the interval ν to $\nu + d\nu$, then

$$E = \int kT\,P\!\left(\frac{h\nu}{kT}\right) z(\nu) d\nu \tag{2}$$

where the integral is to be taken over the entire range of frequencies. Considering the crystal as an elastic continuum, Debye first found the asymptotic law of distribution

$$z(\nu)d\nu = A\nu^2 d\nu, \tag{3}$$

where A can be calculated from the elastic constants. From this it follows immediately that for low temperatures E becomes proportional to T^4, which agrees with observation.

I shall not further develop the theory here, but only explain the more fundamental causes, such as I see them, for the frequency distribution in the crystal. Let us consider the oscillations in a simple lattice, the equations of which have the form

$$m\ddot{u}_x^l + S \sum_{l'\,y} \Phi_{xy}^{l-l'}\,u_y^{l'} = 0, \tag{4}$$

where $\Phi_{xy}^{l-l'}$ is the second derivative of the potential energy at that point of the lattice which is at the distance $\mathbf{r}^{ll'} = \mathbf{r}^{l'} - \mathbf{r}^l$

THE LATTICE THEORY OF RIGID BODIES   193

from the lattice point l. In reality the boundary points of the finite crystal satisfy somewhat different equations, and it is therefore impossible, if the boundary is arbitrary, to reach definite results. Our method then consists in replacing the true boundary conditions by practically equivalent fictitious ones. Imagine a portion of the crystal which is similar to the individual cells, but which contains $N = L^3$ of them. Then we can consider the crystal as infinite, and limit ourselves to those disturbances u_x^l which are periodic with respect to the big cell L^3, that is, those for which

$$u_x^{l+L} = u_x^l.$$

The equation of motion can easily be solved for such a "cyclic" crystal. Set

$$u_x^l = u_x e^{i(\phi l) - i\omega t}$$
$$(\phi l) = \phi_1 l_1 + \phi_2 l_2 + \phi_3 l_3$$
$$\phi_1 = \frac{2\pi}{L} p_1, \cdots$$

where p_1, p_2, p_3 take all integral values from 0 to $(L-1)$. The condition of periodicity, that is, our fictitious boundary condition, is thus fulfilled, and the equations of motion become

$$-m\omega^2 u_x + \Sigma_y a_{xy} u_y = 0 \qquad (5)$$

where
$$a_{xy} = S \Phi_{xy}^l e^{-i(\phi l)},$$

the sum being taken over the whole infinite lattice. Thus the problem is reduced to one of three degrees of freedom.

The frequencies are found as roots of the determinantal equation:

$$\begin{vmatrix} a_{xx} - m\omega^2 & a_{xy} & a_{xz} \\ a_{yx} & a_{yy} - m\omega^2 & a_{yz} \\ a_{zx} & a_{zy} & a_{zz} - m\omega^2 \end{vmatrix} = 0. \qquad (6)$$

To each set of three values p_1, p_2, p_3 belong three roots ω^2. The distribution of the roots is seen most clearly if we consider a cube of side 2π in the ϕ_1, ϕ_2, ϕ_3-space, and in this cube all points whose coördinates are multiples of the Lth part of the side. To each of these $L^3 = N$ points correspond three proper frequencies, which remains true however great the value of L. We can therefore state that in the ϕ-space the proper frequencies are uniformly distributed, so that in the interval $d\phi = d\phi_1 d\phi_2 d\phi_3$, there are $dz = \dfrac{3\,N}{(2\,\pi)^3}\,d\phi$ proper frequencies. This is the basis of the distribution law. If we do not have a simple lattice, but one consisting of s simple lattices, the factor 3 must be replaced by 3 s.

Debye's law is obtained if the calculations are performed in terms of $d\nu$ instead of $d\phi$, which of course can be done only approximately if we do not have an exact solution of the determinantal equation (of the $3s$-th degree).

Debye's approximation can be found in the following way. It is easily seen that our solution represents a plane wave. If we place

$$\phi_1 = \frac{2\,\pi}{\lambda}\mathbf{a}_1 \cdot \mathbf{s},\ \phi_2 = \frac{2\,\pi}{\lambda}\mathbf{a}_2 \cdot \mathbf{s},\ \phi_3 = \frac{2\,\pi}{\lambda}\mathbf{a}_3 \cdot \mathbf{s},\ |\mathbf{s}| = 1,$$

then the exponent becomes

$$(\phi l) - \omega t = \frac{2\,\pi}{\lambda}(\mathbf{r}^l \cdot \mathbf{s}) - \omega t,\ \mathbf{r}^l = l_1 \mathbf{a}_1 + l_2 \mathbf{a}_2 + l_3 \mathbf{a}_3.$$

Therefore λ is the wave-length and $\mathbf{s}$ a unit vector normal to the wave-front. $\dfrac{2\,\pi}{\lambda}\mathbf{s}$ can be considered as the polar coordinate in a space where the rectangular coördinates are

$$\frac{2\,\pi}{\lambda}s_x,\quad \frac{2\,\pi}{\lambda}s_y,\quad \frac{2\,\pi}{\lambda}s_z.$$

Then the relation given above between ϕ_1, ϕ_2, ϕ_3 and these coördinates is a linear transformation of determinant Δ. Therefore

$$d\phi = d\phi_1 d\phi_2 d\phi_3 = \Delta d\left(\frac{2\,\pi}{\lambda}s_x\right)d\left(\frac{2\,\pi}{\lambda}s_y\right)d\left(\frac{2\,\pi}{\lambda}s_z\right)$$

$$= (2\,\pi)^3 \Delta \frac{1}{\lambda^2} d\left(\frac{1}{\lambda}\right) d\Omega$$

where $d\Omega$ is the surface element of a unit sphere.

THE LATTICE THEORY OF RIGID BODIES 195

We shall assume with Debye that the velocity of sound $c = \lambda\nu$ is approximately independent of λ and depends only on the direction of the wave. This assumption is, to a certain degree, fulfilled for regular or almost regular monatomic crystals (simple lattices). Therefore, Debye's theory has been most successful when applied to these monatomic crystals. For other bodies purely mathematical difficulties enter, of which we shall not speak. Because of the hypothesis concerning the speed of sound, $d\phi$ can be integrated over a sphere. Introducing a "mean velocity of sound" $\bar{c}$ defined by

$$\frac{1}{\bar{c}^3} = \int \frac{1}{c^3} \frac{d\Omega}{4\pi}$$

we obtain

$$dz = \frac{3N}{(2\pi)^3} \int \frac{d\phi}{d\Omega} d\Omega = \frac{12\pi V}{\bar{c}^3} \nu^2 d\nu \tag{7}$$

where
$$V = N\Delta$$

is the volume of the crystal. We have thus found Debye's relation and the meaning of the constants it contains.

By a more exact estimate of the different forms of oscillation in a crystal (calculation of dz in the ν-scale) the formulas developed by various investigators are obtained. These express satisfactorily the variation of specific heat with temperature even for complicated and strongly anisotropic crystals. We should also call attention here to some recent work by Grüneisen on the strongly anisotropic crystals of the metals zinc, cadmium and mercury. Grüneisen has also furthered the lattice theory by experimental researches on the relations between mechanical and thermal processes (thermal expansion). This can be only briefly mentioned here.

LECTURE 10

Thermal expansion and pyroelectricity — Concluding remarks.

The atomic theory of thermal expansion and related effects was originated by Debye. He compared this process to the oscillations of an anharmonic oscillator, for which the law of force is asymmetric with respect to the origin. Then the mean position of the oscillating particle is farther from the origin, the greater the amplitude. It is easily shown that this effect is proportional to the mean energy. A similar condition exists in a crystal, where the internal forces oppose the approach of atoms more than their recession. Therefore, the mean distance increases with increasing energy of oscillation, approximately proportionally to this energy, and the effect depends on the deviation from Hooke's law (proportionality between force and displacement). Consider the lattice not in its natural equilibrium, but in a homogeneously disturbed state defined by the deformations u_1, u_2, $\cdots u_{xx} \cdots u_{yz} \cdots$. The thermal effect can then be calculated formally by considering the oscillations of the atoms about their new positions. These oscillations can be considered as approximately harmonic, but due to the deviation from Hooke's law, all the frequencies ν of the lattice, and the thermal energy E as well, depend on the deformation quantities u_k, u_{xy}. Instead of E it is better to consider Helmholtz's "free energy" F as a function of the deformation quantities and the temperature. It can be obtained from the free energy of a resonator, which according to Planck is $kT \ln \left(1 - e^{\frac{-h\nu}{kT}}\right)$, by a summation over all the proper frequencies. To do this the expression above must be multiplied by $z(\nu)$ and integrated over ν. The energy of deformation Φ of the lattice must be added, and we obtain

$$F = \Phi + \int kT \ln \left(1 - e^{\frac{-h\nu}{kT}}\right) z(\nu) d\nu. \tag{1}$$

THE LATTICE THEORY OF RIGID BODIES

Here Φ as well as the integral is a function of the disturbances $\mathbf{u}_k$ and u_{xy}. The integral depends moreover on the temperature.

The inner forces and strains are given, according to thermodynamics, by the formulas

$$K_{kx} = -\frac{\partial F}{\partial u_{kx}}$$

$$K_{xy} = -\frac{\partial F}{\partial u_{xy}}.$$

We shall apply this to the special case of cubic lattices. The function Φ has been given before. It has no linear, but only quadratic terms in the components of the deformation. The frequencies ν of the disturbed lattice will depend linearly on the components of the disturbance, that is only on the combination

$$u_{xx} + u_{yy} + u_{zz} = x_x + y_y + z_z$$

which represents a relative change of volume. Quadratic terms in $\mathbf{u}_k$, u_{xy} also enter which depend on the temperature. We have therefore

$$F = f(x_x + y_y + z_z) + \frac{A}{2}(x_x^2 + y_y^2 + z_z^2) + B(y_y z_z + z_z x_x + x_x y_y)$$
$$+ \frac{B}{2}(y_z^2 + z_x^2 + x_y^2)$$
$$+ C\{(u_{1x} - u_{2x})y_z + (u_{1y} - u_{2y})z_x + (u_{1z} - u_{2z})x_y\}$$
$$+ \frac{D}{2}(\mathbf{u}_1 - \mathbf{u}_2)^2. \qquad (2)$$

A, B, C, D are now functions of the temperature which, with decreasing temperature, approach the constant values which have been considered before, and the new constant f is a function of the temperature which vanishes for $T \to 0$.

Instead of Equation (7) Lecture 3 we now have

$$\begin{aligned}
-K_{1x} &= +K_{2x} = Cy_z + D(u_{1x} - u_{2x}), \cdots \\
-X_x &= f + Ax_x + B(y_y + z_z), \cdots \\
-Y_z &= By_z + C(u_{1x} - u_{2x}) \cdots.
\end{aligned} \qquad (3)$$

If the inner stresses disappear the shearing strains $y_z \cdots$ vanish, but not the components x_x, y_y, z_z. The latter satisfy the equation

$$-f = A x_x + B(y_y + z_z), \cdots.$$

By addition we find

$$x_x + y_y + z_z = \frac{-3f}{A + 2B} = 3\alpha.$$

3α, then, is the fractional increase in volume. Further,

$$x_x = -\frac{f}{A + 2B} = \alpha. \tag{4}$$

α is therefore the coefficient of linear expansion. To calculate α it is necessary to study the frequencies ν in a disturbed lattice. Brody and myself have developed the formula for the most general case, and have shown that in this case as well the assumption of electrostatic cohesion gives a result of the correct order of magnitude.

The formulas for anisotropic crystals show that the linear expansion in the directions parallel and perpendicular to the crystal axis are given by

$$\alpha_{||} = \gamma_{11} p_{||} + \gamma_{12} p_\perp$$
$$\alpha_\perp = \gamma_{21} p_{||} + \gamma_{22} p_\perp$$

where $p_{||}$, $p_\perp$ are the "thermal pressures" in the two perpendicular directions. These are given by the statistical theory as temperature functions

$$p_{||} = f_1(T), \qquad p_\perp = f_2(T)$$

similar to those given by Debye and the constants γ are the elastic moduli. Here $\gamma_{12} = \gamma_{21}$ is negative, for it corresponds to the lateral contraction for a longitudinal force ($p_\perp = 0$).

The character of the functions $f_1(T)$ and $f_2(T)$ depends on the proper frequencies of the lattice. $f_1(T)$ depends chiefly on the oscillations in the direction of the principal axis of the crystal, $f_2(T)$ on those perpendicular to this axis. If the crystal is strongly anisotropic, the frequencies of vibration perpendicular and parallel to the principal axis may be quite different.

The smaller the value of ν, the lower the temperature at which Planck's function $P\left(\dfrac{h\nu}{kT}\right)$ begins to increase rapidly with T, and the functions f_1 and f_2 have the same property. It is therefore possible that at low temperatures f_1 should have a sensible value while f_2 is still vanishingly small. Then, γ_{21} being neg-

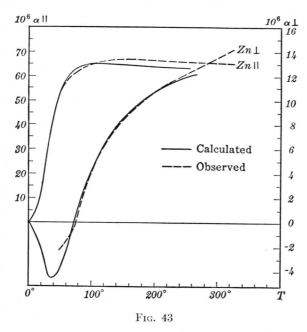

Fig. 43

ative, it follows that for low temperatures $\alpha_\perp$ will be negative, and will only later become positive through the influence of $f_2(T)$.

This peculiar phenomenon of a negative coefficient of linear expansion in certain directions in anisotropic crystals was first discovered for zinc and cadmium by Grüneisen, who gave the above explanation. From measurements of the constants of elasticity he showed that the vibrations parallel and perpendicular to the principal axis have very different frequencies, and was able, by approximate calculations of the constants γ and of the functions $f_1(T)$ and $f_2(T)$, to represent the variation of the coefficient of expansion in good agreement with experiment.

For crystals of lesser symmetry, linear terms in the components of the inner displacement u_k also enter in the expression of the free energy F. The equilibrium conditions then give relative displacements of the simple lattices for vanishing stresses, with coefficients which depend on the temperature.

For ion lattices an electric moment which depends on the temperature is therefore obtained, i.e., the phenomenon of pyroelectricity. The variation of the latter with temperature should again be similar to that of the specific heat. Experiments of Ackermann have shown an analogous decrease of the pyroelectric coefficients with decreasing temperature. He found at the lowest temperatures a linear variation with T, while the theory demands proportionality to T^3. According to Heckmann this is probably to be explained in a manner similar to the negative coefficient of expansion of Grüneisen. Several functions of the temperature which oppose each other give this apparent linear relation. Crucial measurements and calculations have not yet been made.

In this brief survey I have only been able to give a small part of the work done on the dynamics of crystal lattices, but I hope that it will suffice to give you a picture of the bases and objects of the theory. If you ask what I think will be the future development of the theory I shall answer that, for ion lattices, there remains little more to be done. It seems to me that progress in the knowledge of lattice structure is only possible on the basis of a better knowledge of the structure of atoms and molecules. As long as it is not known how the hydrogen atom is built, it seems to me that extensive researches on the structure of the diamond or other non-polar crystals will not be very fruitful. I may, therefore, give my opinion that a concentration of the forces of all experimental and theoretical physicists on the problems of atomic structure and quantum theory is the next step in order to extend our knowledge of crystals.